PROGRAMMABLE CONTROLLERS

HARDWARE
SOFTWARE
AND
APPLICATIONS

GEORGE L. BATTEN, JR., Ph.D.

TAB Professional and Reference Books

Division of TAB BOOKS Inc.
Blue Ridge Summit, PA

Dedication
For Jeanne, Brigid, Katie, and Jason

Trademarks

LoLa™ Programming Language is a trademark of Furnas Electric Co.

NumeriExpress™ Communications Link is a trademark of Giddings & Lewis Electronics Company.

Series One™ Programmable Controllers, Series Three™ Programmable Controllers, Series Six™ Programmable Controllers, and Workmaster™ Programmable Controller Information Center are trademarks of GE Fanuc Automation Company

TPR books are published by TAB Professional and Reference Books, a division of TAB BOOKS Inc. The TPR logo, consisting of the letters ''TPR'' within a large ''T,'' is a registered trademark of TAB BOOKS Inc.

FIRST EDITION
SECOND PRINTING

Library of Congress Cataloging in Publication Data

Batten, Jr., George L.
   Programmable controllers : hardware, software, and applications /
by George L. Batten, Jr.
   p. cm.
   Bibliography: p.
   Includes index.
   ISBN 0-8306-3147-X
   1. Programmable controllers. I. Title.
TJ223.P76B38 1988
629.8'95—dc19
                                      88-25034
                                         CIP

TAB BOOKS Inc. offers software for sale. For information and a catalog, please contact TAB Software Department, Blue Ridge Summit, PA 17294-0850.

Questions regarding the content of this book should be addressed to:
Reader Inquiry Branch
TAB BOOKS Inc.
Blue Ridge Summit, PA 17294-0214

Contents

Introduction

In 1968, THE STATE OF THE ART IN CONTROL TECHNOLOGY WAS THE RELAY CONTROL SYStem. Shortly thereafter, the programmable controller became the state of the art in control technology. Since then, programmable controllers have made rapid inroads into virtually every manufacturing business and many service businesses. The relatively rapid integration of the programmable controller into the manufacturing sector has been called "the second industrial revolution." And the revolution is far from over.

It is obvious from the foregoing that the successful technician, engineering aid, engineer, or supervisor involved in nearly any manufacturing business needs a basic knowledge of the way in which programmable controllers are used. In response to this demand for knowledge about programmable controllers, several good books have appeared. These books tend to go into great detail concerning programmable controller hardware, criteria for selection, and maintenance. This is information that the working engineer needs to know. Unfortunately, the first-time reader of these books is lost because each of these books assumes some previous knowledge of programmable controllers on the part of the reader. Thus technicians, engineering aids, engineers, and supervisors who need an introduction to programmable controllers are confronted with a dilemma: The acquisition of basic knowledge about programmable controllers depends upon the existence of a basic knowledge about programmable controllers. Where does the uninitiated reader turn for this basic knowledge?

This book is the answer to that question. Simply stated, the purpose of this book is to provide the reader with a basic understanding of programmable controllers.

Let us be specific. A reader, previously unfamiliar with programmable controllers, should be able to do the following, upon completion of this book: (1) understand the basic components of programmable controller systems; (2) have a reasonable understanding of the languages used to program these devices; (3) understand how programmable controllers are used in closed-loop, feedback control schemes (especially proportional-integral-derivative control); (4) understand how programmable controllers communicate with peripheral devices and other programmable controllers in networks; (5) be able to read a technical data sheet on a programmable controller and understand the capabilities and limitations contained therein, without being overwhelmed by the sales pitch; and (6) be able to pick up any of the advanced texts listed in the bibliography and benefit from reading them. In other words, this book will not turn the previously uninitiated reader into an expert on programmable controllers, but it will give the reader a solid foundation on which to build expertise.

The purpose of the book thus defines the audience of the book. It is meant for technical people (technicians, engineering aids, and engineers) and supervisory personnel who require a basic understanding of programmable controller operations and uses. You do not need to have any special knowledge, skill, or experience. Some basic understanding of electricity/electronics is assumed. If you understand how a relay works, your level of electricity/electronics education is sufficient. And if you have a passing familiarity with personal computers, this book will be a breeze.

The book divides naturally into sections. The first section includes Chapters 1 through 5 and provides background material. Chapter 1 defines the programmable controller and compares programmable controllers with personal computers. The history of the programmable controller is recounted, and an incomplete list of industries employing programmable controllers is given. Chapter 2 introduces *logic circuits*. These are the circuits that form the building blocks of the programmable controller's central processing unit. The logic operations introduced in this chapter also appear later in the form of a programming language. Chapter 3 introduces Boolean algebra, the natural mathematics of the circuits introduced in Chapter 2. Chapter 4 deals yet again with numbers. This time, the focus is number systems. Our own decimal number system is compared with systems that find widespread use in programmable controllers (and personal computers). These are the *binary, octal,* and *hexadecimal* number systems. Since the programmable controller's microprocessor actually uses the binary system, that system is highlighted. Binary arithmetic is developed, and circuits for the performance of binary arithmetic are described. Finally, binary number codes are introduced. Chapter 5 examines the relay and how it was used in control applications. Since the programmable controller was developed to replace relay control systems, this chapter is essential to an understanding of programmable controllers. The ubiquitous *relay-ladder diagram* makes its first appearance in this chapter.

Chapters 6, 7, and 8 form the second section of the book. This section addresses programmable controller hardware. Chapter 6 covers the central processing unit: the power supply, microprocessor, and memories. Chapter 7 looks at input/output interfaces, the devices that allow the programmable controller to adapt to the outside world. This chapter also contains a thorough discussion of proportional, proportional-integral, and proportional-integral-derivative control. And Chapter 8 covers peripherals, or external devices (such as printers) that are connected to the programmable controller.

Chapter 9 forms the third section of the book. If you own a personal computer, you know that it is of very little use without appropriate software, or programs. The business of programming a programmable controller is complicated by two factors: there are a variety of programming languages to choose from, and the most popular ones are based on the logic used to wire relay control systems. Relay-ladder diagrams are a holdover from the days before programmable controllers, and tend to confuse younger engineers who never used relays in control. High-level computer languages tend to confuse the older engineers, who still remember the days when relays controlled plant processes. Both low-level (relay-type) and high-level (computer-type) programming languages are described in this chapter. A comparison of both types of languages points out the similarities (and disparities) between the two.

The fourth section of the book is Chapter 10, and the fifth section is Chapter 11. Chapter 10 provides practical examples of how programmable controllers are used in process measurement and control, and Chapter 11 describes how they communicate with each other in local area networks. These chapters round out the basic introduction to programmable controllers.

Section Six includes Chapter 12, the Glossary, the Bibliography, and the Appendices. This is a section of reference material. Chapter 12 provides an overview of the programmable controllers available in the United States at the time of this writing. In light of the information contained in the first eleven chapters of the book, it is interesting to note the diversity of programmable controllers and systems. The glossary is convenient for quick reference. The bibliography suggests texts for further reading, as well as courses and periodicals.

The appendices provide a variety of material: (1) names and addresses of suppliers of programmable controllers, (2) introductory information on a new, high-level programming language, (3) a summary of Boolean algebra, (4) the ASCII code, (5) a summary of relay-ladder and Boolean programming symbols, and (6) selected literature on commercial programmable controllers. The suppliers' literature included in Appendix F has been selected to illustrate not only the full-sized programmable controllers used for large control schemes, but also the low-end (small, inexpensive) programmable controllers that are causing quite a stir these days.

Now you know why the book was written, who should read it, and what to expect. It is time now to forge ahead and to begin learning about these remarkable, indispensable devices. Good luck!

Chapter 1

Background

WHAT IS A PROGRAMMABLE CONTROLLER? BRIEFLY, IT IS A DIGITAL ELECTRONIC DEVICE that meets the following three criteria:

- It has a programmable memory, in which instructions can be stored
- The instructions stored in the memory are used to implement various functions, such as logic, sequencing, timing, counting, and arithmetical functions
- The various functions are used to control machines or processes

The control functions of the programmable controller (PC) are accomplished through input/output modules, which can be either analog or digital.

From this description, it would seem that a personal computer can be used as a programmable controller. Indeed, personal computers *can* be used as programmable controllers, as shown in Fig. 1-1.

In Fig. 1-1 a personal computer is used as a home-security programmable controller. In this example, three types of sensing elements are connected to the personal computer via an input interface module: a photocell which measures light level, a smoke detector (or several smoke detectors), and various security switches on doors and windows. The personal computer's memory has been programmed with instructions to be implemented by the microprocessor. For example, if the light level in a room falls below a predetermined value, the microprocessor is instructed to turn on the room lights. If a signal is received from the smoke detector, the microprocessor is instructed to sound an alarm and to dial the fire department and deliver a prerecorded message. By performing these specified functions, the personal computer controls light switches, burglar and fire alarms, and telephone communications via the output interface module. Thus the personal computer in this application satisfies the three criteria listed above, and functions as a programmable controller.

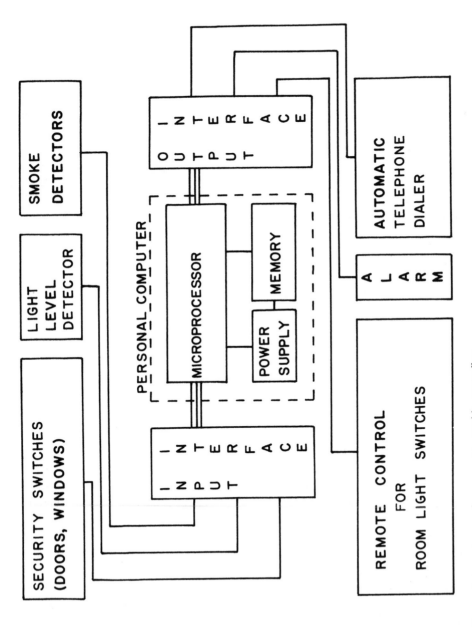

Fig. 1-1. Personal computer used as a programmable controller.

Although personal computers can function as programmable controllers, they are generally not used in that application. There are several reasons for this. Personal computers are extremely flexible devices that can be used to perform calculations, to plot graphs, to play games, and to instruct children. Why should one tie up such a flexible and useful device for the mundane purpose of home security? Why not develop a microprocessor-based home-security programmable controller that is designed to perform the one series of tasks only? This is one of the principal reasons why personal computers are not generally used as programmable controllers.

A second reason why personal computers are not used as programmable controllers involves the environment in which the control functions are performed. Many manufacturing plants do not have carefully controlled temperatures and humidities. Additionally, plants that use pumps and motors generally experience electrical interference problems. Personal computers are not designed to perform in such rugged environments, but programmable controllers are so designed.

In summary, the programmable controller shares many features with the personal computer: it is a digital, electronic, programmable device that functions to control machines or processes. The programmable controller differs from the personal computer in that the PC is more rugged and somewhat less flexible than the personal computer.

THE HISTORY OF THE PROGRAMMABLE CONTROLLER

The history of the programmable controller is not extensive. In fact, it goes back only as far as 1968, to the Hydramatic Division of General Motors Corporation.

The mass manufacture of automobiles (or, for that matter, automobile components) involves many machines, all of which must be controlled. Consider, for example, a boring machine. If the boring machine is used to bore a hole in the center of a piece of metal, then there must be some control mechanism that prevents the boring machine from operating until the piece of metal is aligned properly. Also, this control mechanism must actuate the boring machine when the metal is aligned and reverse the direction of the bit when boring is completed. Prior to 1968, this control function was performed by control relays. (The way in which relays are used as control devices will be described in a later chapter.)

Control relays were effective, but they suffered from several disadvantages. To begin with, relays are only capable of on/off control, so in order to design complicated control systems, many relays are needed. This makes the relay control scheme quite expensive. Control relays can be bulky, so a control system requiring many relays takes up much floor and cabinet space. Control relays are power-hungry, and this high power consumption results in heat generation. When a relay fails, either through an opening of the coil or a pitting of the contacts, it is difficult to troubleshoot and locate the failed relay. But worst of all, relays are hardwired. Any change in the control program requires the rewiring of relays. This is extremely costly, both in terms of labor and plant downtime, and (in the case of General Motors) is of critical importance, since relay control systems must be changed each year due to model year differences.

Thus, General Motors had sufficient incentive to eliminate the relay-control system. In 1968, GM specified design criteria for a programmable-logic controller (PLC) to replace the relay-control system. These criteria required that the PLC must be easily programmed or reprogrammed, with a minimum of downtime (or loss of service); must be easily maintained (i.e., that it must be modular and self-diagnosing); must be rugged enough to operate in the industrial environment; must consume less power and require less cabinet and floor space than the relay control system; and must be competitive in cost. Additional specifications required that the PLC have expandable memory, communicate with data collection systems, and accept 120-volt ac signals.

This specification attracted the attention of several manufacturers of control equipment, and the results of their efforts were the first-generation programmable controllers. Those first programmable controllers were quite primitive, when evaluated by today's standards. They, like the relays they replaced, simply functioned as on/off controllers. However, the reduced space requirements and power-consumption specifications were met, and the PCs did have primitive self-diagnostic indicators that aided in troubleshooting. These features caused the primitive PCs to be widely used, and with widespread use came improvements. Also, developments in microprocessor technology translated into more flexible and powerful PCs.

As time progressed, PCs acquired the capacity to do arithmetic, to manipulate data, and to communicate more efficiently with the programmer. Following innovations in microprocessor and memory technology, PCs have acquired larger memory capacity and the ability to communicate with other controllers or a master *host* computer. With the development of analog control, PCs were able to move past simple on/off control to more complex schemes, such as proportional-integral-derivative (PID) control. And, with these developments, the cost of PCs decreased steadily.

In summary, today's PCs are economical, user-friendly devices that have benefitted from advances in the microprocessor and memory technologies. They are capable of performing complicated control routines and can communicate with other PCs and host computers in sophisticated control networks.

PC APPLICATIONS

There are very few industries today that do not employ programmable controllers. Almost every business in the manufacturing sector, and many in the service sector, use PCs in abundance. A very incomplete list of industries that use PCs would include the following: aerospace, automotive, bottling and canning, chemicals, entertainment, food and beverage, gas and petroleum, lumber, machining, metals, mining, packaging, petrochemicals, plastics, power, pulp and paper, rubber, and transportation. In these industries, PCs perform a variety of control functions from weighing, conveying, and handling materials to drilling, boring, converting, and packaging.

It is little wonder that the relatively rapid integration of the PC into the manufacturing sector has been called the second industrial revolution. And the revolution is far from being finished.

Chapter 2

Introduction to Logic Circuits

CHAPTER 1 PRESENTED THREE CRITERIA THAT PROGRAMMABLE CONTROLLERS MUST MEET: the presence of a programmable memory for the storage of instructions; the use of the instructions to implement various functions; and the use of the functions to control machines or processes. These criteria require the operation of logic circuits. In this chapter the elements of circuit logic are outlined, and the basic logic circuits, such as the AND, OR, and NOT gates, are discussed.

THE AND GATE

If you rent a safety deposit box at the local bank, you are familiar with the AND gate. In order to enter a safety deposit box, two keys are required: the renter's key *and* the banker's key. If both keys are inserted and turned at the same time, the safety deposit box is opened.

The electrical equivalent of this is the AND function; it is shown in Fig. 2-1. In this simple model the function to be performed is that of turning on the lamp. This occurs when two conditions are satisfied simultaneously; i.e., when switch number 1 and switch number 2 are both closed.

A variation on the circuit in Fig. 2-1 can be created using relays. This is shown in Fig. 2-2. In this figure, the light is turned on when both relays are energized simultaneously; i.e., when a signal or voltage is applied to both relays. Thus the AND function is performed only when a voltage is applied to relay number 1 *and* relay number 2.

The circuit in Fig. 2-2 is shown redrawn in Fig. 2-3. In this figure, the relays and battery have been enclosed by a dashed line, which is the symbol for an AND gate. Modern semiconductor AND gates are not composed of relays, of course, but the principle of operation is the same as for the relay circuit shown in Fig. 2-3: when voltages are applied to all inputs of the AND gate, an output voltage is obtained.

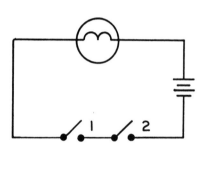

Fig. 2-1. Simple model of an AND gate.

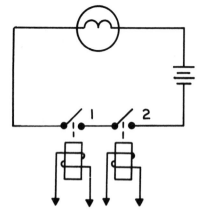

Fig. 2-2. Fig. 2-1 redrawn with relays.

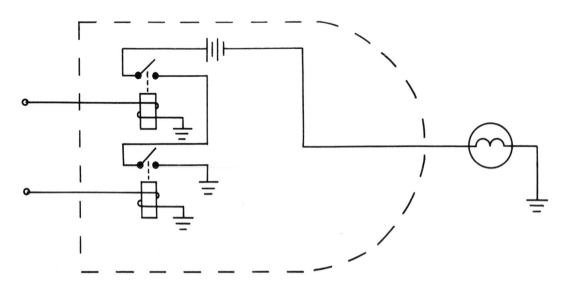

Fig. 2-3. Fig. 2-2 redrawn.

The modern semiconductor AND gate is a *binary* device. This means that it recognizes only two states of existence: off or on. (A more accurate statement would be that the device recognizes low or high voltages. Digital logic circuits generally use + 5 volts dc as the on level and 0 volts dc as the off level.) The off state (or low state, or false state) is symbolized by the number 0, and the on state (or high state, or true state) is symbolized by the number 1. Thus if any input to the AND gate is 0, then the output is 0; if all inputs are 1, the output is 1. (Remember that 0 and 1 represent off and on, respectively, and the number 1 usually corresponds to a voltage of + 5 volts dc.)

The AND gate is not limited to two inputs. Three or more inputs to an AND gate are not uncommon. Two-input AND gates can be used to build larger input AND gates; i.e., four-, six-, or eight- input AND gates, as shown in Fig. 2-4.

A complete description of the outputs of an AND gate for possible sets of inputs can be represented conveniently with a *truth table*. This is shown in Fig. 2-5 for a three-input AND gate. Each horizontal line of the truth table lists a possible combination of inputs and the output for that combination of inputs. For example, the first line of the

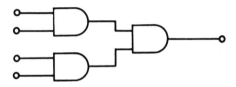

Fig. 2-4. A four-input AND gate constructed from three two-input AND gates.

AND Truth Table

	Inputs		Output
A	B	C	Y
0	0	0	0
0	0	1	0
0	1	0	0
1	0	0	0
1	1	0	0
0	1	1	0
1	0	1	0
1	1	1	1

Fig. 2-5. Truth table of a three-input AND gate.

truth table shows that, for three false inputs, the output is false. Scanning down the table, one sees that the output is true *only* when *all* the inputs to the AND gate are true.

THE OR GATE

A simple model for the OR gate is shown in Fig. 2-6. When either switch number 1 *or* switch number 2 is closed, the lamp is turned on. This simple model, like that of the AND gate, can be modified to include the use of relays, as is shown in Fig. 2-7. In this figure, the lamp is activated if a signal is supplied either to relay number 1 *or* relay number 2.

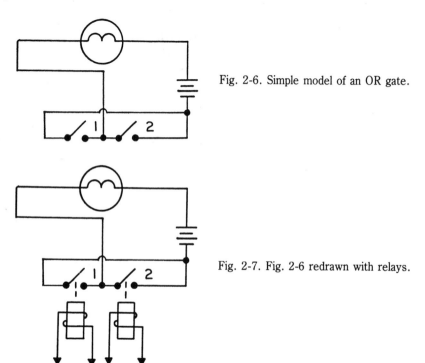

Fig. 2-6. Simple model of an OR gate.

Fig. 2-7. Fig. 2-6 redrawn with relays.

The symbol for an OR gate is shown in Fig. 2-8, along with its truth table. The only set of inputs that produces a false or 0 output is the first set; i.e., the set composed of all 0 inputs.

The OR gate is not limited to two inputs. OR gates with three or more inputs are not uncommon. As with the AND gate, two-input OR gates can be used to build larger input OR gates, as shown in Fig. 2-9.

THE NOT GATE

A simple model for the NOT gate is shown in Fig. 2-10. The function performed by the circuit is that of supplying an ac voltage to the lamp. The relay used in the circuit is normally closed. When an on signal is supplied to the relay, the contact opens and the lamp switches to the off state. The output of the circuit is the opposite, or inverse, of the input to the relay.

The symbol for the NOT gate is shown in Fig. 2-11, along with its truth table. The truth table is simple: the output is always the inverse of the input. If the input is 1, the output is NOT 1 (0), and vice versa. That is why the NOT gate is often called an *inverter*.

```
OR Truth Table


     Inputs          Output


    A    B              Y

    0    0              0

    0    1              1

    1    0              1

    1    1              1
```

Fig. 2-8. Truth table of a two-input OR gate.

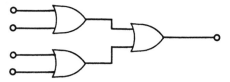

Fig. 2-9. A four-input OR gate constructed from three two-input OR gates.

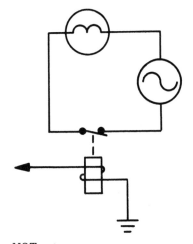

Fig. 2-10. Simple model of a NOT gate.

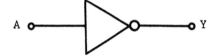

NOT Truth Table

Input	Output
A	Y
0	1
1	0

Fig. 2-11. Truth table of a NOT gate.

OTHER LOGIC GATES

Three other logic gates deserve mention. These gates, the NAND, NOR, and EXCLUSIVE-OR (XOR) gates, are *composite* gates; that is, they are composed of combinations of AND, OR, or NOT gates.

The principle of operation of the NAND gate, as well as its symbol and truth table, is shown in Fig. 2-12. The NAND gate is essentially an AND gate followed by an

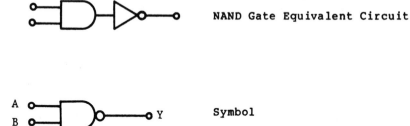

NAND Gate Equivalent Circuit

Symbol

Truth Table

Inputs		Output
A	B	Y
0	0	1
0	1	1
1	0	1
1	1	0

Fig. 2-12. The NAND gate.

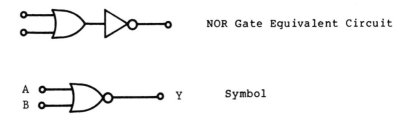

NOR Gate Equivalent Circuit

Symbol

Truth Table

Inputs		Output
A	B	Y
0	0	1
0	1	0
1	0	0
1	1	0

Fig. 2-13. The NOR gate.

inverter—NOT AND, or NAND. If the output column of the AND gate truth table is inverted, one obtains the NAND gate truth table. In other words, the output of the NAND gate is 1 if *any* input is 0. If all inputs are 1, then the output is 0.

The NOR gate is similar in construction. It is essentially an OR gate followed by a NOT gate. Inverting the output column of the OR gate truth table yields the NOR gate truth table, as shown in Fig. 2-13. The output of the NOR gate is 1 *only* when *all* inputs to the NOR gate are 0.

The EXCLUSIVE-OR (XOR) gate is more complicated than the NAND and NOR gates. It is the equivalent of one OR, two AND, and two NOT gates connected as shown in Fig. 2-14. The only difference between the OR and XOR gates can be seen in the last line of the truth table in Fig. 2-14. While the OR gate gives an output of 1 when all inputs are 1, the XOR gate gives an output of 0 when all inputs are 1.

POSITIVE AND NEGATIVE LOGIC

In all the examples given above, the 1 state has been represented with a more positive voltage than that representing the 0 state. For example, it was stated previously that the 1 state usually corresponds to a voltage of + 5 volts dc, while the 0 state usually corresponds to a voltage of 0 volts dc. This type of logic, in which the 1 state is represented by a more positive voltage than the 0 state, is called *positive logic*.

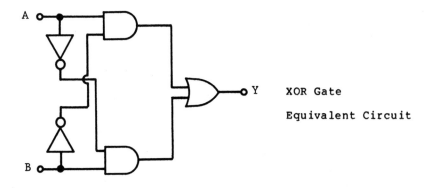

XOR Gate

Equivalent Circuit

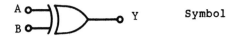

Symbol

Truth Table

Inputs		Output
A	B	Y
0	0	0
0	1	1
1	0	1
1	1	0

Fig. 2-14. The EXCLUSIVE-OR (XOR) gate.

There is a second type of binary logic, called *negative logic*. With negative logic, the 1 state is represented by the less positive voltage. For example, in a negative logic device, the 1 state might correspond to 0 volts dc, while the 0 state would correspond to + 5 volts dc.

Usually, positive logic is the prevalent logic; however, sometimes it is more convenient to work with negative logic. You should be aware of its existence.

SUMMARY

The logic gates described in this chapter perform basic logic functions and are the building blocks of PCs. In later chapters, these gates will be used to perform control functions, which range from the simple to the complex.

Chapter 3

Boolean Algebra

MATHEMATICIANS ARE NOTORIOUS FOR INVENTING NEW FORMS OF MATHEMATICS THAT have no apparent practical applications, but that later prove to be useful. George Boole (1815-1864), the English mathematician and logician, was no exception. He developed an algebra of sets that bears his name, *Boolean algebra*. The symbolism of set theory is confusing to most nonmathematicians, so a complete definition of Boolean algebra, including its postulates, will not be attempted here. It will be sufficient to note that Boolean algebra provides a convenient shorthand for describing logic operations, particularly those operations of the logic gates discussed in the previous chapter. This chapter serves as an introduction to the elementary principles of Boolean algebra.

BOOLEAN EXPRESSIONS

In the last chapter, the letters *A, B, C,* and so on were used to symbolize inputs to the logic gates, while the letter *Y* was used to designate the output of the logic gates. That convention will be maintained in this chapter.

There are three basic Boolean expressions of interest to those who work with logic circuits: the multiplication sign, the addition sign, and the horizontal bar above a letter, or combination of letters.

The multiplication sign ($\times$, $\cdot$, or letters written together with nothing between them) is a symbolic shorthand for the AND operation. Thus the equation describing the three-input AND gate shown in Fig. 3-1 is:

$$Y = A \times B \times C \qquad \text{(3-1a)}$$

or

$$Y = A \cdot B \cdot C \tag{3-1b}$$

or

$$Y = ABC \tag{3-1c}$$

This can be seen clearly by examining the various horizontal rows in the truth table. In the first row, $A=B=C=0$. When multiplied together, the result is $Y=0$. The same is true for the second row; i.e., $0 \times 0 \times 1 = 0$. This holds all the way down the truth table. When any input is zero, the output is zero. The last row of the truth table provides the only nonzero output. When $A=B=C=1$, their product is $Y=1$.

The addition sign is a symbolic shorthand for the OR operation. Thus the equation describing the two-input OR gate shown in Fig. 3-2 is:

$$Y = A + B \tag{3-2}$$

Again, this can be seen by examining the rows in the truth table. If $A=B=0$, then $A+B=0$. For any nonzero input, $A+B=1$. And of course, when both inputs are 1, the output is 2. In short, if the sum of the inputs is 1 (or more than 1), the output is 1.

The horizontal bar above a letter is a symbolic shorthand for the NOT operation. Thus the equation describing the inverter shown in Fig. 3-3 is:

$$Y = \overline{A} \tag{3-3}$$

This is shown in the truth table. Y is always NOT A, or $\overline{A}$.

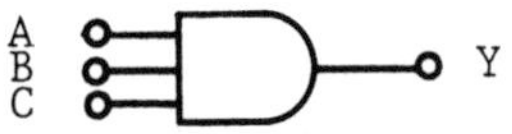

AND Truth Table

Inputs			Output
A	B	C	Y
0	0	0	0
0	0	1	0
0	1	0	0
1	0	0	0
1	1	0	0
0	1	1	0
1	0	1	0
1	1	1	1

Fig. 3-1. Truth table of a three-input AND gate.

OR Truth Table

Inputs		Output
A	B	Y
0	0	0
0	1	1
1	0	1
1	1	1

Fig. 3-2. Truth table of a two-input OR gate.

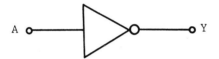

NOT Truth Table

Input	Output
A	Y
0	1
1	0

Fig. 3-3. Truth table of a NOT gate.

BOOLEAN ALGEBRA

In the previous chapter it was shown how AND, OR, and NOT gates could be combined to form NAND, NOR, and XOR gates. The Boolean expressions presented in the previous section can also be combined to describe these composite gates. The rules for combining Boolean expressions are given in this section.

As with "ordinary" algebra, Boolean algebra obeys the two following commutative laws:

$$A + B = B + A \qquad (3\text{-}4)$$

$$AB = BA \qquad (3\text{-}5)$$

In other words, it does not matter in which order two inputs are added (ORed) or in which order two inputs are multiplied (ANDed). Also, Boolean algebra obeys the two following associative laws:

$$A + (B+C) = (A+B)+C \tag{3-6}$$

$$A(BC) = (AB)C \tag{3-7}$$

This simply means that, when dealing with more than two inputs, the grouping of the inputs to either an OR or an AND gate does not matter.

The commutative and associative laws given above describe "pure" logic operations; i.e., purely AND operations or purely OR operations. Mixing of these operations is covered by the distributive law:

$$A(B+C) = AB+AC \tag{3-8}$$

While this law is the same as for "ordinary" algebra, it is worthwhile to read Equation 3-8, substituting the words "and" and "or" for the multiplication and addition signs, respectively: "A and (B or C) equals A and B, or A and C." This approach emphasizes the fact that Boolean algebra is simply a mathematical way of stating a logical sentence.

There is another distributive law that is best demonstrated with logic gates. Consider the following equation:

$$Y = A + BC \tag{3-9}$$

The logic circuit and truth table for this equation are given in Fig. 3-4. From the truth table it is seen that $Y=1$ when $A=1$, or when *both* B and $C=1$. Now consider the following equation:

$$Y = (A+B)(A+C) \tag{3-10}$$

The logic circuit and truth table for this equation are given in Fig. 3-5. The truth tables for these two circuits are identical. Since the truth tables are identical, and since Equations 3-9 and 3-10 are simply Boolean equations that can be used to generate the truth table, then the two equations must be equal. In other words, we have another distributive law:

$$A+BC = (A+B)(A+C) \tag{3-11}$$

Note that this is not the result one would obtain with "ordinary" algebra. If one followed the normal rules of algebra, expanding the right hand side of Equation 3-11 would result in:

$$(A+B)(A+C) = AA+AC+BA+BC \tag{3-12}$$

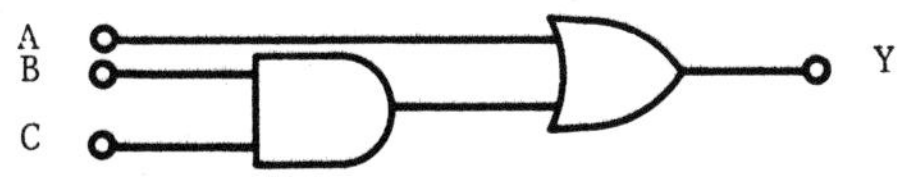

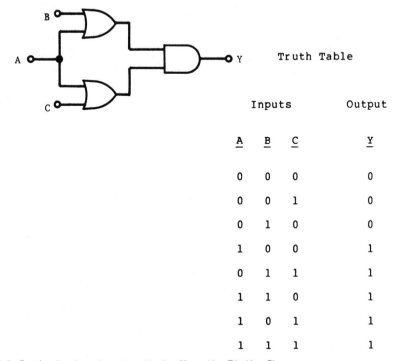

Truth Table

	Inputs		Output
A	**B**	**C**	**Y**
0	0	0	0
0	0	1	0
0	1	0	0
1	0	0	1
0	1	1	1
1	1	0	1
1	0	1	1
1	1	1	1

Fig. 3-4. Logic circuit and truth table for Y = A + BC.

Truth Table

	Inputs		Output
A	**B**	**C**	**Y**
0	0	0	0
0	0	1	0
0	1	0	0
1	0	0	1
0	1	1	1
1	1	0	1
1	0	1	1
1	1	1	1

Fig. 3-5. Logic circuit and truth table for Y = (A + B) (A + C).

This expanded equation, however, does yield the same truth table found in Figs. 3-4 and 3-5. Obviously, if $A = 1$ then $AA = 1$, and if $A = 0$ then $AA = 0$. So in Boolean algebra, $AA = A$. Using this identity and the commutative laws (Equations 3-4 and 3-5), Equation 3-12 may be rewritten as:

$$(A+B)\ (A+C) = A + AB + AC + BC \qquad (3\text{-}13)$$

Using the distributive law, Equation 3-8, the first three terms on the right-hand side can be rewritten as one term, $A(1+B+C)$:

$$(A+B)\ (A+C) = A\ (1+B+C) + BC \qquad (3\text{-}14)$$

Now consider the term $A\ (1+B+C)$. It really doesn't matter what the values of B or C are, since there is a one in the parentheses. The value of the term $A(1+B+C)$ will always be determined by the value of A: If A is one, the term will have the value of 1. If A is zero, the term will have the value of zero. Thus we can replace $A(1+B+C)$ with A, and Equation 3-14 becomes:

$$(A+B)\ (A+C) = A + BC \qquad (3\text{-}15)$$

which is just the same as Equation 3-11.

In the preceding derivation of Equation 3-15, two identities were discovered: Any input multiplied ("ANDed") by itself is just the input; and any input multiplied (ANDed) by a term of added (ORed) inputs that includes the number 1 is just the original input. In other words,

$$AA = A \qquad (3\text{-}16)$$

$$X(1+A+B+C+...) = X \qquad (3\text{-}17)$$

These two identities can be combined to yield the law of absorption:

$$A(A+B) = A \qquad (3\text{-}18)$$

If we expand the left-hand side of Equation 3-18, we have:

$$A(A+B) = AA + AB \qquad (3\text{-}19)$$

Application of Equation 3-16 yields:

$$A(A+B) = A + AB = A(1+B) \qquad (3\text{-}20)$$

The application of Equation 3-17 to Equation 3-20 results in Equation 3-18.

The laws of Boolean algebra governing inversion deserve special mention. Figure 3-6 gives the logic circuit and truth table for $Y = \overline{(A+B)}$. Note that the operation in parentheses, *A OR B*, is performed first, followed by the inversion. The intermediate output, Y', gives the output one would expect from the OR gate; however, the final output Y is the inverted (NOTed) value of Y'. Now consider Fig. 3-7, which gives the logic circuit and truth table for $Y = \overline{A}\overline{B}$. Note that the NOT operation is performed prior to performing the AND operation. The two truth tables are the same. This means that the two Boolean equations are identical:

$$\overline{(A+B)} = \overline{A}\overline{B} \qquad (3\text{-}21)$$

This identity is known as one of *DeMorgan's Laws*.

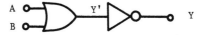

Truth Table

Inputs		Outputs	
A	**B**	**Y'**	**Y**
0	0	0	1
0	1	1	0
1	0	1	0
1	1	1	0

Fig. 3-6. Logic circuit and truth table for $Y = \overline{(A+B)}$.

Another of DeMorgan's laws can be derived fairly easily. Figure 3-8 shows the logic circuits and truth table for $Y = \overline{(AB)}$ and $Y = \overline{A}+\overline{B}$. Again, both circuits follow the same truth table, so they must be equal. This yields another of DeMorgan's laws:

$$\overline{(AB)} = \overline{A}+\overline{B} \qquad (3\text{-}22)$$

The order in which Boolean operations are performed is of extreme importance. Generally, operations in parentheses are to be performed first. Thus in the left-hand side of Equation 3-22, the AND operation is performed prior to the NOT operation. When

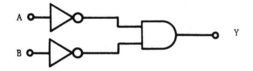

Truth Table

	Inputs		Output
A	**B**		**Y**
0	0		1
0	1		0
1	0		0
1	1		0

Fig. 3-7. Logic circuit and truth table for $Y = \overline{A}\,\overline{B}$.

$Y = \overline{(AB)}$

$Y = \overline{A} + \overline{B}$

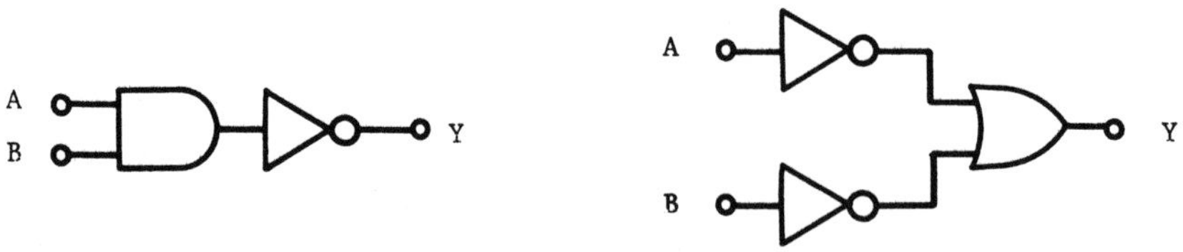

Truth Table

	Inputs		Output
A	**B**		**Y**
0	0		1
0	1		1
1	0		1
1	1		0

Fig. 3-8. Demonstration of DeMorgan's law.

no parentheses are present, the priority order for operations are NOT, AND, and then OR. Thus in the right-hand side of Equation 3-22, the NOT operation is performed on both inputs prior to the OR operation.

When dealing with inversion, be careful to note the extent of the horizontal line above the symbols. It is left as an exercise for you to demonstrate that $(\overline{AB})$ is *not* equal to $\overline{A}\overline{B}$.

In the preceding derivations of DeMorgan's laws, Equations 3-21 and 3-22, only two inputs were considered. It can be shown that DeMorgan's laws hold for more than two inputs:

$$\overline{(A+B+C+D+E)} = \overline{A}\,\overline{B}\,\overline{C}\,\overline{D}\,\overline{E} \tag{3-23}$$

$$\overline{(ABCDE)} = \overline{A}+\overline{B}+\overline{C}+\overline{D}+\overline{E} \tag{3-24}$$

When an input is inverted twice, its value is just that which it had prior to inversion:

$$\overline{\overline{A}} = A \tag{3-25}$$

where the double bar represents inversion twice. Obviously, this holds true for AND and OR operations on inputs:

$$\overline{\overline{(AB)}} = AB \tag{3-26}$$

$$\overline{\overline{(A+B)}} = A+B \tag{3-27}$$

DeMorgan's laws can also be applied to twice-inverted inputs:

$$\overline{(\overline{A}+B+C)} = \overline{\overline{A}}\,\overline{B}\,\overline{C} = A\overline{B}\,\overline{C} \tag{3-28}$$

I close this section with three useful identities. Consider Boolean Equation (3-29):

$$Y = AB + A\overline{B} \tag{3-29}$$

Using the distributive law, it can be written as

$$Y = A(B+\overline{B}) \tag{3-30}$$

If B is 0, then $\overline{B}$ is 1, and vice versa, so that the value of Y will be determined by the value of A. In other words,

$$AB + A\overline{B} = A \tag{3-31}$$

The second identity is derived as follows. Consider Equation 3-32:

$$Y = A + \overline{A}B \tag{3-32}$$

If $A=1$, then $Y=1$. If $A=0$, then $\overline{A}=1$ and the value of Y will be determined by the value of B. In other words, Y will equal either A or B. Thus,

$$A+\overline{A}B \;=\; A+B \tag{3-33}$$

For the third identity, consider Equation 3-34 :

$$Y \;=\; AB+AC+B\overline{C} \tag{3-34}$$

The truth table for this Boolean equation is as follows:

A	B	C	Y
0	0	0	0
0	0	1	0
0	1	0	1
1	0	0	0
0	1	1	0
1	1	0	1
1	0	1	1
1	1	1	1

Now consider Equation 3-35:

$$Y \;=\; AC+B\overline{C} \tag{3-35}$$

The truth table for this equation is the same as for Equation 3-34, so the two equations must be equal. This yields the third identity:

$$AB+AC+B\overline{C} \;=\; AC+B\overline{C} \tag{3-36}$$

BOOLEAN EXPRESSIONS FOR COMPOSITE GATES

As discussed in the previous chapter, there are three composite gates of interest: the NAND, NOR, and XOR gates. The Boolean expressions and identities presented thus far can be used to describe these composite gates.

The NAND gate (Fig. 3-9) is an AND gate followed by a NOT gate. Its Boolean equation is thus:

$$Y \;=\; \overline{(AB)} \tag{3-37}$$

From DeMorgan's law, Equation 3-22, it can be seen that the NAND gate is also described by

$$Y \;=\; \overline{A}+\overline{B} \tag{3-38}$$

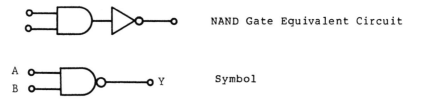

NAND Gate Equivalent Circuit

Symbol

Truth Table

Inputs		Output
A	B	Y
0	0	1
0	1	1
1	0	1
1	1	0

Fig. 3-9. The NAND gate.

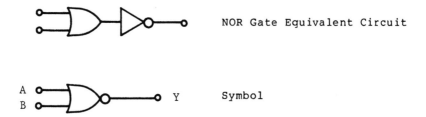

NOR Gate Equivalent Circuit

Symbol

Truth Table

Inputs		Output
A	B	Y
0	0	1
0	1	0
1	0	0
1	1	0

Fig. 3-10. The NOR gate.

The NOR gate (Fig. 3-10) is an OR gate followed by a NOT gate. Its Boolean equation is thus:

$$Y = \overline{(A+B)} \tag{3-39}$$

From DeMorgan's law, Equation 3-21, it can be seen that the NOR gate is also described by:

$$Y = \overline{A}\,\overline{B} \tag{3-40}$$

The EXCLUSIVE-OR (XOR) gate (Fig. 3-11) is complicated when compared with the NAND and NOR gates, but its equation is straightforward:

$$Y = (A\overline{B}) + (\overline{A}B) \tag{3-41}$$

A comparison of Equations 3-37 through 3-41 with the truth tables in Figs. 3-9 through 3-11 shows that the Boolean equations yield the appropriate truth tables.

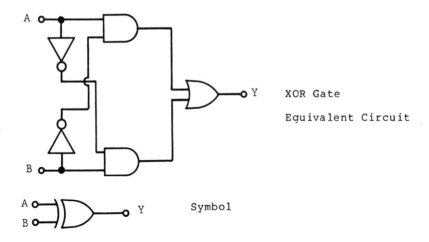

XOR Gate

Equivalent Circuit

Symbol

Truth Table

Inputs		Output
A	B	Y
0	0	0
0	1	1
1	0	1
1	1	0

Fig. 3-11. The EXCLUSIVE-OR (XOR) gate.

SUMMARY

Boolean algebra provides a convenient shorthand for describing logic operations, particularly those operations of the logic gates discussed in the previous chapter. A summary of the laws and identities of Boolean algebra follows:

Operations

$$\times \;=\; \text{AND}$$
$$+ \;=\; \text{OR}$$
$$\overline{A} \;=\; \text{NOT } A$$

Laws

Commutative

$$A+B \;=\; B+A$$
$$AB \;=\; BA$$

Associative

$$A+(B+C) \;=\; (A+B)+C$$
$$A(BC) \;=\; (AB)C$$

Distributive

$$A(B+C) \;=\; AB+AC$$
$$A+BC \;=\; (A+B)\,(A+C)$$

Absorptive

$$A(A+B) \;=\; A$$

DeMorgan's

$$\overline{(A+B+C)} \;=\; \overline{A}\,\overline{B}\,\overline{C}$$

$$\overline{(ABC)} \;=\; \overline{A}+\overline{B}+\overline{C}$$

Identities

$$AA \;=\; A$$
$$X(1+A+B+C+D+...) \;=\; X$$
$$\overline{\overline{A}} \;=\; A$$
$$\overline{\overline{(AB)}} \;=\; AB$$

$$\overline{\overline{(A+B)}} = A+B$$

$$AB+A\overline{B} = A$$

$$A+\overline{A}B = A+B$$

$$AB+AC+B\overline{C} = AC+B\overline{C}$$

Chapter 4

Computer Arithmetic and Coding

THE LOGIC GATES THAT FORM THE HEART OF THE PROGRAMMABLE CONTROLLER (PC) ARE binary devices, so some familiarity with the binary number system is required of PC users. In addition, other number systems and codes are often used. These include the octal and hexadecimal number systems, and the Binary-Coded Decimal (BCD), Gray, Baudot, and ASCII Codes. This chapter reviews these number systems and codes.

THE DECIMAL NUMBER SYSTEM

The *decimal* number system is presented first, since it is the one with which we are most familiar. In this number system, the *base*, or *radix*, is ten, which is also the number of digits in this system: 0 through 9. With these ten digits, numbers of any magnitude can be generated as powers of ten.

In order to show this, we introduce *scientific notation*. Scientific notation is generally used to represent numbers that are either very large or very small, but in this case, it is a convenient way of demonstrating the decimal number system. Scientific notation involves a base (or radix) and an exponent, or superscript. Since we are dealing with the decimal system, the base is ten. The exponent tells how many times the base is involved in self-multiplication. For example, 10^1 is just 10, while 10^2 is 10×10 or 100. Likewise, 10^3 is $10 \times 10 \times 10$, or 1000, and 10^4 is $10 \times 10 \times 10 \times 10$, or 10,000, etc. There is a zeroth power of ten. *Any* number raised to the zeroth power is just 1. Thus, $10^0 = 1$.

Consider now the number one thousand two hundred thirty four (1,234). This number is the sum of 1 thousand, 2 hundreds, 3 tens, and 4 ones. In other words

$$1{,}234 = 1 \times 10^3 + 2 \times 10^2 + 3 \times 10^1 + 4 \times 10^0 \qquad (4\text{-}1)$$

Each *digit* in the number occupies a *position* that has a certain *weight*. The position farthest to the right (digit = 4) has the weight 10^0, or 1. The position second from the right (digit = 3) has the weight 10^1, or 10. This continues to the left, and can be made to extend to infinity. Another example is presented in Fig. 4-1.

As can be seen from Fig. 4-2, decimal numbers can also be represented in this way. Positions to the right of the decimal place have negative exponential weights, such as 10^{-1} (= 0.1) and 10^{-2} (= 0.01).

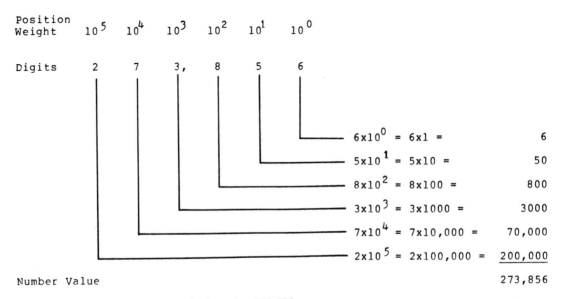

Fig. 4-1. Representation of the decimal number 273,856.

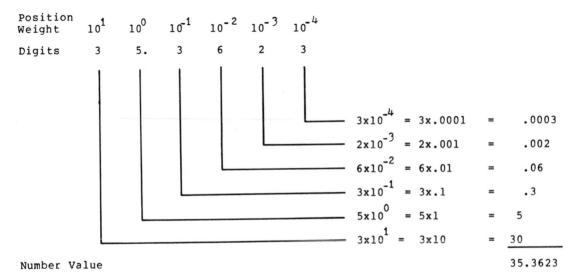

Fig. 4-2. Representation of the decimal number 35.3623.

Since we are very familiar with the decimal number system, we do not stop to consider the weight of each position. Automatically we recognize that the number 432 represents 4 hundreds, 3 tens, and 2 ones. When dealing with other number systems (i.e., systems with bases other than ten), however, it will be necessary to evaluate numbers with the method shown in Fig. 4-1. The next section will illustrate this.

THE BINARY NUMBER SYSTEM

The binary number system has a base of two, which is also the number of digits in this system: 0 and 1. With these two digits, numbers of any magnitude can be generated as powers of two.

Each digit in a number occupies a position that has a certain weight. This can be demonstrated using Fig. 4-3, which shows the binary number 11011. The position farthest to the right (digit = 1) has the weight 2^0, or 1. (Remember: any number raised to the zeroth power is just 1.) The second position from the right (digit = 1) has the weight 2^1, or 2. The third position from the right (digit = 0) has the weight 2^2, or 4. The fourth position from the right (digit = 1) has the weight 2^3, or 8. And the furthermost left position (digit = 1) has the weight 2^4, or 16. Thus the binary number 11011 is the equivalent to the decimal number 27:

$$11011 = 1 \times 2^4 + 1 \times 2^3 + 0 \times 2^2 + 1 \times 2^1 + 1 \times 2^0$$
$$11011 = 1 \times 16 + 1 \times 8 + 0 \times 4 + 1 \times 2 + 1 \times 1 \tag{4-2}$$
$$11011 = 16 + 8 + 2 + 1 = 27$$

The advantages of using the binary number system should be obvious. Logic gates exist in one of two possible states: either on or off, 1 or 0. A number system that has only two digits is a natural number system for digital logic operations. Thus the binary

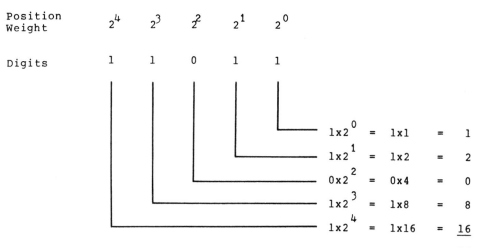

Fig. 4-3. Representation of the binary number 11011.

number system is the primary language of logic circuits, computers, and programmable controllers.

Converting Decimal Numbers to Binary Numbers

It is easy to convert a number from the binary to the decimal number system. You simply use the method outlined above to obtain Equation 4-2. In other words, each digit in the binary number is multiplied by its weighting factor, and the total is summed. Converting a number from the decimal to the binary number system is not quite as straightforward, but it is not terribly complicated. The method for doing this is often called *the dibble-dabble method*; it is demonstrated in Fig. 4-4.

DECIMAL NUMBER 79

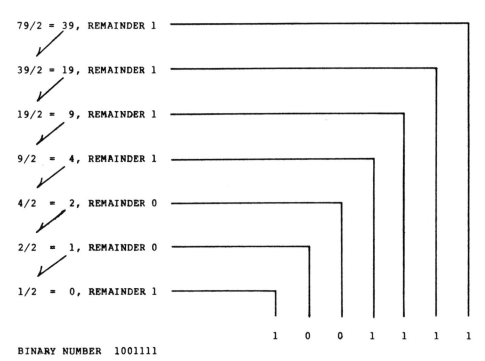

```
79/2 =  39, REMAINDER 1
39/2 =  19, REMAINDER 1
19/2 =   9, REMAINDER 1
9/2  =   4, REMAINDER 1
4/2  =   2, REMAINDER 0
2/2  =   1, REMAINDER 0
1/2  =   0, REMAINDER 1

         1   0   0   1   1   1   1
```

BINARY NUMBER 1001111

Fig. 4-4. The dibble-dabble method for converting decimal numbers to binary numbers.

In this figure we are converting the decimal number 79 to its binary equivalent. The first step involves dividing 79 by 2:

$$79/2 = 39, \text{ with a remainder of } 1$$

The remainder of 1 becomes the first digit in the binary number, i.e., the digit with a weight of 2^0, or 1. The second digit is obtained by dividing 39 by 2:

$$39/2 = 19, \text{ with a remainder of } 1$$

The remainder of 1 becomes the second digit in the binary number, the digit with a weight of 2^1, or 2. We continue on in this fashion:

$$
\begin{aligned}
79/2 &= 39, \text{ remainder } 1 \\
39/2 &= 19, \text{ remainder } 1 \\
19/2 &= 9, \text{ remainder } 1 \\
9/2 &= 4, \text{ remainder } 1 \\
4/2 &= 2, \text{ remainder } 0 \\
2/2 &= 1, \text{ remainder } 0 \\
1/2 &= 0, \text{ remainder } 1
\end{aligned}
$$

When read from bottom to top, the remainders give us the binary equivalent of decimal 79, or 1001111.

It is easy to double-check this answer. We simply convert the binary number 1001111 to decimal, and check to see that the answer is 79:

$$
\begin{aligned}
1001111 &= 1\times2^6+0\times2^5+0\times2^4+1\times2^3+1\times2^2+1\times2^1+1\times2^0 \quad (4\text{-}3)\\
1001111 &= 1\times64+0\times32+0\times16+1\times8+1\times4+1\times2+1\times1 \\
1001111 &= 64+8+4+2+1 \\
1001111 &= 79
\end{aligned}
$$

The dibble-dabble method is thus an effective method of converting a number from the decimal number system to the binary number system.

Binary Arithmetic

We are all familiar with the four basic arithmetical operations in the decimal number system: addition, subtraction, multiplication, and division. These operations can also be performed in the binary number system.

The addition table for the binary number system is quite simple:

$$
\begin{aligned}
0+0 &= 0 \quad &(4\text{-}4)\\
0+1 &= 1 \quad &(4\text{-}5)\\
1+1 &= 10 \quad &(4\text{-}6)\\
10+1 &= 11 \quad &(4\text{-}7)
\end{aligned}
$$

To see how this table is used in the addition of binary numbers, let us add two numbers:

$$
\begin{array}{r}
1011 \\
+1100 \\
\hline
\end{array}
$$

We begin, as in the decimal number system, with the column of numbers farthest to the right (i.e., the *least significant digit*) and proceed to the left-most column of numbers (i.e., the *most significant digit*). Equation 4-5 allows us to fill in the three columns to the right, while Equation 4-6 allows us to fill in the column of most significant digits:

$$
\begin{array}{r}
1011 \\
+1100 \\
\hline
10111
\end{array}
$$

The result above can be checked by converting all the numbers to the decimal system. The binary number 1011 is the decimal number 11. Binary number 1100 is decimal number 12. Their sum is 23. In the binary number system, 23 is represented as 10111. Thus the result checks out.

Numbers are "carried over" to the next column in binary addition just as in decimal addition. Addition of the binary numbers 10111 and 11111 provides a convenient example. Beginning with the least significant digits, $1+1=10$, we record the zero, and carry over the one. The next column is $1+1=10$, plus the carried-over one, which yields a total of 11. We record the least significant one and carry over a one. This continues to the column of most significant digits:

$$
\begin{array}{r}
10111 \\
+11111 \\
\hline
110110
\end{array}
$$

To check this solution, we convert to the decimal number system. Binary 10111 equals decimal 23, and binary 11111 equals decimal 31. Their sum is decimal 54, which is binary 110110. The solution checks out.

Binary addition is easily accomplished with logic circuits, as will be seen in the next section.

The rules of subtraction in the binary number system are as follows:

$$ 0-0 \;=\; 0 \tag{4-8} $$

$$ 1-0 \;=\; 1 \tag{4-9} $$

$$ 1-1 \;=\; 0 \tag{4-10} $$

$$ 10-1 \;=\; 1 \tag{4-11} $$

As an example, consider subtracting 10 from 11:

$$
\begin{array}{r}
11 \\
-10 \\
\hline
\end{array}
$$

Equation 4-9 gives us the solution for the least significant digit, while Equation 4-10 gives us the solution for the most significant digit:

$$
\begin{array}{r}
11 \text{ (minuend)} \\
-10 \text{ (subtrahend)} \\
\hline
1 \text{ (difference)}
\end{array}
$$

where the zero in the most significant digit column is not written. As a check, binary 11 equals decimal 3, binary 10 equals decimal 2, and their difference is 1, in either number system.

Since 0-1 is not an allowed operation in binary subtraction, when 0 appears in the minuend above a subtrahend digit of 1, we must "borrow" from the next digit to the left, just as in subtraction in the decimal number system. An example should make this clear. Assume that we wish to subtract 489 from 537 in the decimal number system:

column	(a)	(b)	(c)
	5	3	7
	−4	8	9

We begin with column c. Obviously, 7 is less than 9, so the 7 "borrows" a 1 (actually a 10) from column b:

column	(a)	(b)	(c)
	5	3	(17)
	−4	8	9
			8

The subtraction, 17−9, is performed, and we are finished with column c.

In column b, the 1 borrowed by column c must be repaid, but *it is repaid to the subtrahend*. Thus the 8 in the subtrahend gains a 1, and becomes a 9:

column	(a)	(b)	(c)
	5	3	(17)
	−4	(9)	9
			8

Now 3 is less than 9, so it borrows a 1 from column a.

column	(a)	(b)	(c)
	5	(13)	(17)
	−4	(9)	9
		4	8

The subtraction 13-9 completes our concern with column b. We move to column a and repay the borrowed 1 to the 4 in the subtrahend:

column	(a)	(b)	(c)
	5	(13)	(17)
	− (5)	− (9)	9
	0	4	8

We can drop the zero to the left (column a), and see that the result of 537 − 489 is 48. Adding the difference (48) to the subtrahend (489) results in 537, which is the minuend. Thus the result checks out.

The same procedure is followed in binary subtraction. Consider the problem of 11010 minus 1111:

column	(a)	(b)	(c)	(d)	(e)
	1	1	0	1	0
−		1	1	1	1

We begin with column e. In column e the minuend borrows a 1 from column d:

column	(a)	(b)	(c)	(d)	(e)
	1	1	0	1	(10)
−		1	1	1	1
					1

wherein we have used Equation 4-11. This 1 gets repaid to the subtrahend in column d.

column	(a)	(b)	(c)	(d)	(e)
	1	1	0	1	(10)
−		1	1	(10)	1
					1

Column d now borrows a 1

column	(a)	(b)	(c)	(d)	(e)
	1	1	0	(11)	(10)
−		1	1	(10)	1
				1	1

which is paid back to the subtrahend in column c:

column	(a)	(b)	(c)	(d)	(e)
	1	1	0	(11)	(10)
−		1	(10)	(10)	1
				1	1

The minuend in column c now borrows a 1

column	(a)	(b)	(c)	(d)	(e)
	1	1	(10)	(11)	(10)
−		1	(10)	(10)	1
			0	1	1

which gets paid back to column b of the subtrahend:

column	(a)	(b)	(c)	(d)	(e)
	1	1	(10)	(11)	(10)
−		(10)	(10)	(10)	1
			0	1	1

The process is repeated to completion:

column	(a)	(b)	(c)	(d)	(e)
	1	(11)	(10)	(11)	(10)
−(1)	(10)	(10)	(10)	(10)	1
	0	1	0	1	1

Dropping the left-most zero yields the result.

$$\begin{array}{r} 11010 \\ -1111 \\ \hline 1011 \end{array}$$

This result can be checked in two ways. First, we add the difference to the subtrahend. The result should be the minuend.

$$\begin{array}{r} 1011 \\ +1111 \\ \hline 11010 \end{array}$$

Second, we convert to decimal numbers and perform the subtraction:

$$
\begin{array}{rcr}
11010 & = & 26 \\
-\ 1111 & = & -15 \\
\hline
1011 & = & \overline{11}
\end{array}
$$

The subtraction checks both ways.

In the decimal number system, subtraction can be viewed as the addition of a negative number to a positive number. Logic circuits, computers, and PCs do not recognize negative numbers; thus subtractions performed by electronic logic devices cannot be carried out using the method described above. Instead, the method of *complements* is used by computers and PCs to perform binary subtraction.

There are two types of complements: the *one's complement* and the *two's complement*. We consider the one's complement first.

In the one's complement method an extra digit is placed in the leftmost column of the number, the most significant digit column. This digit is 0 if the number is positive, and 1 if the number is negative. If the number is negative, each digit in the number is inverted (i.e., 1s become 0s, and 0s become 1s). For example, the binary number 1011 is equal to the decimal number 11. We now give the number a *sign* digit: 0 is positive, 1 is negative:

$$\text{binary } 01011 = \text{decimal } +11$$

To represent the decimal number -11 in the binary system, we invert *all* the digits:

$$\text{binary } 10100 = \text{decimal } -11$$

What is the result when we add decimal numbers $+11$ and -11? The result is zero, of course. In the binary number system, we have:

$$
\begin{array}{r}
01011 \\
+10100 \\
\hline
11111
\end{array}
$$

which is not zero. In order to obtain zero, we must add a 1 to the least significant digit, and ignore the most significant digit:

$$
\begin{array}{r}
01011 \\
+10100 \\
+\quad\ 1 \\
\hline
100000 = 0, \text{ when MSD is ignored.}
\end{array}
$$

This example outlines how subtraction is performed with one's complements. Find the complement of the subtrahend, add one to it, add it to the minuend, and ignore the most significant digit.

For a second example, let us subtract decimal 22 from decimal 41 (result = 19) in the binary system.

decimal	binary
41	101001
− 22	− 10110
19	10011

To begin, we add the positive sign digit to 101001 and take the complement of 10110, making sure to add the negative sign (1) as the most significant digit.

$$0101001$$
$$101001$$

Next, add 1 to the subtrahend, and add the subtrahend to the minuend:

$$0101001$$
$$101010$$
$$1010011$$

If we ignore the most significant digit (the 1 farthest to the left), the result is 010011, or positive 10011. Thus the solution checks out.

As a final example, we subtract 1011 (decimal number 11) from 1110 (decimal number 14). The result should be 11 (decimal number 3);

1110	→	01110	→	01110
− 1011		+ 10100		+ 10101
				100011 = positive 11

The solution is correct.

The two's complement method is somewhat different from the one's complement method. The extra digit in the most significant column is still used (0 = positive, 1 = negative), and inversion of some digits takes place. However, inversion (from right to left) *occurs only after the first 1 is detected.*

An example should make this clear. Binary number + 1110 (01110, or decimal + 14) has the one's complement of:

Binary number	01110
one's complement	10001

To find the two's complement, we begin with the least significant digit, which is a zero. Since a one has not been detected, the zero is left unchanged. We move to the next column to the left. The digit is a one, the first one detected. It is left unchanged, *but all digits after this are inverted*:

Binary number	01110
two's complement	10010

Thus in two's complement terminology, 10010 is negative decimal 14.

When the binary number is added to the two's complement, the result should be zero. As seen below, it is zero *if the most significant digit is ignored*.

$$
\begin{array}{r}
01110 \\
+\,10010 \\
\hline
100000
\end{array}
$$

This provides us with the rules of subtraction using two's complements. First, find the two's complement of the subtrahend. Second, add it to the minuend. Third, ignore the most significant digit of the sum. The following two examples should make this clear.

Decimal	Binary	Two's Complement
41	101001	0101001
−22	−10110	101010
19	10011	1010011

or positive 10011

Decimal	Binary	Two's Complement
14	1110	01110
−11	−1011	10101
3	11	100011

or positive 11.

The only difference between subtraction with two's complements and subtraction with one's complements is that, with two's complements, it is not necessary to add a 1 to the complement.

Once addition and subtraction are mastered, multiplication and division are performed easily. Multiplication is just multiple addition. For example, 55×12 is nothing more than 55 added to itself twelve times. Division is multiple subtraction. For example, to divide 16 by 8, we subtract 8 as follows:

$$\begin{array}{r} 16 \\ -8 \\ \hline 8 \end{array} \text{ first subtraction}$$

$$\begin{array}{r} -8 \\ \hline 0 \end{array} \text{ second subtraction}$$

to yield a total of two subtractions; thus, 16/8 = 2.

Of course, the same holds true for the binary number system. 1101 × 101 is just 1101 added to itself 101 times:

$$\begin{array}{r} 1101 \\ \times 101 \\ \hline 1101 \\ 1101 \\ \hline 1000001 \end{array}$$

In the same way, 1100 divided by 110 is just the number of times that 110 can be subtracted from 1100.

$$\begin{array}{r} 10 \\ 110\overline{)1100} \\ \underline{110} \\ 0 \end{array}$$

With the method of complements, subtraction can be performed as addition. Multiplication and division are nothing more than multiple additions and subtractions, respectively. Thus all four functions (addition, subtraction, multiplication, and division) can be performed by an adding circuit. Such a circuit will be discussed in the next section.

Binary Adder Circuits

The two basic types of logic circuit adders are the *half-adder* and the *full adder*. The half-adder symbol, circuit, and truth table are shown in Fig. 4-5. The half-adder adds two digits. It accepts two inputs (the digits to be added) and yields two outputs: the "sum," or S, output, and the "carry," or C, output. The circuit is composed of an EXCLUSIVE-OR (XOR) gate and an AND gate. (Recall, from Chapter 2, that the XOR gate is a composite gate, which is composed of two NOT, two AND, and one OR gates. See Fig. 2-14.) The sum output is given by Equation 4-12:

$$S = A\overline{B} + \overline{A}B \qquad (4\text{-}12)$$

while the carry output is given by Equation 4-13:

$$C = AB \qquad (4\text{-}13)$$

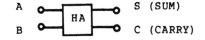

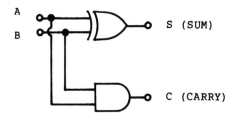

TRUTH   TABLE

INPUTS		OUTPUTS	
A	B	S	C
0	0	0	0
0	1	1	0
1	0	1	0
1	1	0	1

Fig. 4-5. Symbol, circuit, and truth table for the half-adder.

In adding two digits with the half-adder, the operation is performed as follows:

$$\begin{array}{r} A \\ +B \\ \hline CS \end{array}$$

With this in mind, the truth table in Fig. 4-5 yields the following results:

$$\begin{array}{cccc} 0 & 0 & 1 & 1 \\ +0 & +1 & +0 & +1 \\ \hline 00 & 01 & 01 & 10 \end{array}$$

The full adder adds three digits: A, B, and a carry input CI from a previous full adder or half-adder. This is shown in Fig. 4-6. The logic equations for the sum and carry outputs are:

$$S = \overline{A}\,\overline{B}CI + \overline{A}B\overline{CI} + A\overline{B}\,\overline{CI} + ABCI \qquad (4\text{-}14)$$
$$C = AB + (A + B)CI \qquad (4\text{-}15)$$

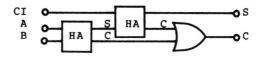

```
              TRUTH   TABLE

          INPUTS        OUTPUTS

        A   B   CI       S   C

        0   0   0        0   0

        0   0   1        1   0

        0   1   0        1   0

        1   0   0        1   0

        0   1   1        0   1

        1   1   0        0   1

        1   0   1        0   1

        1   1   1        1   1
```

Fig. 4-6. Symbol, circuit, and truth table for the full adder.

Using the full adder and half-adder, addition can be performed on numbers containing many digits. A simple example, that of $11+10$, is shown in Fig. 4-7.

Of course, adder circuits can be made quite complex. It is not the purpose of this book to explore the design and construction of adder circuits, but from the examples given above, it can be seen that logic gates can be combined to perform addition, and that addition can be used to perform the other three arithmetical operations: subtraction, multiplication, and division. Multitudes of logic gates can be fabricated on one small integrated circuit chip in such a way as to perform these operations. Such a chip is called a *central processing unit* (CPU), which is the heart of a computer or PC. The CPU is thus nothing more than a collection of logic gates.

Before proceeding to the next section, we introduce some definitions. A *bit* is defined as a digit in the binary number system. The binary number 1011 has four bits, while the binary number 10101010 has eight bits. A *nibble* is a group of four bits. A *byte* is a group of eight bits. A *word* is a group of one or more bytes. With this terminology,

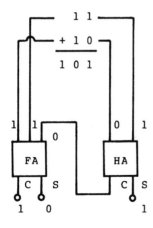

Fig. 4-7. Addition of $11 + 10$ using one full adder and one half-adder.

the most significant digit and least significant digit in the binary number system become the *most significant bit* and *least significant bit*, respectively.

THE OCTAL NUMBER SYSTEM

Since logic circuits, computers, and PCs are binary devices, their primary language is the binary number system. However, other number systems are usually employed for communication with the computer (i.e., input/output addressing) and for data storage (i.e., memory). The two principal systems of communication are the *octal number system* and the *hexadecimal number system*. We discuss the octal number system in this section.

The base, or radix, of the octal number system is 8, which is also the number of digits in this system: 0 through 7. (Also, 8 is the number of bits in a byte.) As before, numbers greater than 7 are represented as powers of 8:

$$
\begin{aligned}
\text{Octal } 1234 &= 1 \times 8^3 + 2 \times 8^2 + 3 \times 8^1 + 4 \times 8^0 \\
&= 1 \times 512 + 2 \times 64 + 3 \times 8 + 4 \times 1 \\
&= 512 + 128 + 24 + 4 \\
&= 668 \text{ decimal}
\end{aligned}
\tag{4-16}
$$

The advantage of the octal system over the binary system is immediately apparent. Four digits are required to represent decimal 668 in the octal system. In the binary system, ten bits are required:

$$\text{binary } 1010011100 = \text{decimal } 668$$

The base of the octal system is 8, which is 2^3. This provides a convenient method of converting between the octal and binary systems. To convert from binary to octal, group the binary number into three-bit groups, beginning with the least significant bit. Then convert into octal by simply converting the groups into decimal. This is shown

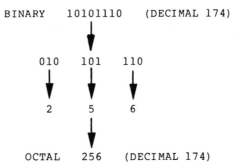

Fig. 4-8. Conversion of a binary number to an octal number.

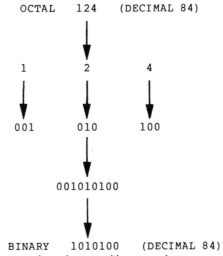

Fig. 4-9. Conversion of an octal number to a binary number.

in Fig. 4-8. To convert from octal to binary, convert each octal digit into a three-bit binary number, beginning with the least significant digit. This is shown in Fig. 4-9.

The easiest way to convert from decimal to octal, and vice versa, is to convert the number first to the binary number system and then use the method outlined in the preceding paragraph.

THE HEXADECIMAL NUMBER SYSTEM

The base, or radix, of the hexadecimal number system is 16, which is also the number of characters in the system: 0 through 9 and A through F. A through F are used to represent numbers 10 through 15 because these numbers have two digits each. Use of two digit numbers would be confusing, and worse, would increase the number of bits necessary to represent bytes or words. Numbers greater than F are represented as powers of 16:

$$
\begin{aligned}
\text{hexadecimal B13} \quad &= 11 \times 16^2 + 1 \times 16^1 + 3 \times 16^0 \\
&= 11 \times 256 + 1 \times 16 + 3 \times 1 \\
&= 2816 + 16 + 3 \\
&= 2835 \text{ decimal}
\end{aligned}
\tag{4-17}
$$

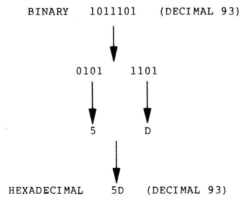

Fig. 4-10. Conversion of a binary number to a hexadecimal number.

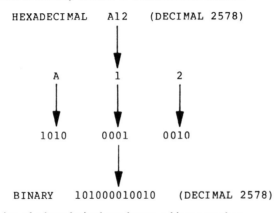

Fig. 4-11. Conversion of a hexadecimal number to a binary number.

The preceding example displays the advantage of using the hexadecimal number system. A number that requires four decimal digits can be represented with only three hexadecimal characters.

Since $2^4 = 16$, each character of a hexadecimal number represents four binary bits. This provides a convenient method for converting back and forth between the binary and hexadecimal number systems, as shown in Figs. 4-10 and 4-11. To convert back and forth between the hexadecimal and octal number systems, or the hexadecimal and decimal number systems, use the binary number system as an intermediary.

The four basic number systems are compared in Table 4-1.

BINARY CODES

A code is just a systematic method for presenting or transmitting information. The binary codes presented in the remainder of this chapter are designed to allow machines to communicate with one another; specifically, to allow the computer to communicate with external devices, such as input/output devices, or *peripherals*.

Table 4-1. Comparison Of The Four Basic Number Systems.

DECIMAL	BINARY	OCTAL	HEXADECIMAL
0	0	0	0
1	1	1	1
2	10	2	2
3	11	3	3
4	100	4	4
5	101	5	5
6	110	6	6
7	111	7	7
8	1000	10	8
9	1001	11	9
10	1010	12	A
11	1011	13	B
12	1100	14	C
13	1101	15	D
14	1110	16	E
15	1111	17	F

Obviously, one of the most important communications links is that of communication between the *operator* (or programmer) and the PC. The codes that follow translate the programmer's language into a form that the PC or computer can use. This is done by assigning to each number, letter, or symbol a unique combination of binary bits.

Of the codes that follow, the Binary Coded Decimal, Gray, and ASCII Codes are quite common. The Baudot Code has fallen into disuse, primarily because of the superiority of the ASCII Code. The ASCII Code is a popular keyboard code (computer and PC terminals). The Baudot Code was used primarily with teleprinter machines.

The Binary Coded Decimal Code

The great advantage of the Binary Coded Decimal (BCD) code is its simplicity. The BCD code converts decimal numbers to 4-bit binary numbers. Thus the decimal numbers 0 through 9 are represented by BCD 0000 through 1001. This is shown in Table 4-2.

Conversion between decimal and BCD is easier than between decimal and binary. For example, the decimal number 23 can be represented in binary as 10111. In order to convert, one must recall the powers of 2 up to the fourth power. The number 23 in BCD is 0010 0011, where the first four bits represent the 2 and the second four bits represent the 3. One needs only recall the powers of 2 up to the third power. Unfortunately, BCD requires more binary bits than the binary system to represent the same number (in this example, 8 in BCD, 5 in binary).

The largest number that can be represented with four binary bits is 1111 (decimal 15). The largest required for BCD is 1001 (decimal 9). Thus the numbers 1010, 1011,

Table 4-2. BCD Equivalents of Decimal Numbers.

DECIMAL	BCD
0	0000
1	0001
2	0010
3	0011
4	0100
5	0101
6	0110
7	0111
8	1000
9	1001

1100, 1101, 1110, and 1111 are not used in BCD. However, these numbers are not wasted as they can be used for sending instructions. For example, 1110 could be sent as an instruction to a stamping machine that tells the machine to change the die pattern.

An excellent example of an application of the BCD code is the pocket calculator. Each number key, when pushed, transmits a BCD signal to the calculator's CPU. For example, when the number 7 key is pressed, 0111 is sent to the calculator's internal digital circuits. Likewise, when the answer is determined and transmitted to the calculator's display, the BCD is converted to decimal for use by the seven-segment LED display.

The Gray Code

In the BCD code, decimal number 1 is represented as 0001, while decimal number 2 is represented as 0010. In order to make the ascent from decimal 1 to decimal 2, two bits must change: the least significant bit changes from 1 to 0, and the bit to the left changes from 0 to 1. The *Gray code (cyclic code,* or *reflected binary code)* requires only one bit to change between successive numbers. The Gray code is presented in Table 4-3.

The Gray code is not suitable for arithmetic operations. Its primary use is in mechanical-to-electrical conversions, such as in positioning applications.

The Baudot Code

The Baudot code found its greatest application in teleprinter machines where, in addition to 10 digits, 26 letters of the alphabet and numerous symbols (such as #, ?, /, ( ), etc.) needed to be represented.

The Baudot code is a *5-level* code, which means that each character on the keyboard (i.e., each character of transmitted data) is represented by 5 binary bits. For example, when the letter F on the teleprinter machine is pressed, the machine transmits the binary code 10110. By using a shift key, the 64 keyboard characters could be represented with the thirty-two Baudot numbers (00000 through 11111).

Table 4-3. The Gray Code.

DECIMAL	GRAY CODE
0	0000
1	0001
2	0011
3	0010
4	0110
5	0111
6	0101
7	0100
8	1100
9	1101
10	1111
11	1110
12	1010
13	1011
14	1001
15	1000

There were two principal drawbacks to the Baudot code: it contained only 32 numbers, and these numbers were not in ascending binary order. The ASCII code, described in the next section, effectively eliminated these problems.

The ASCII Code

ASCII ("as-key") is an acronym for *American Standard Code for Information Interchange*. With this code, 6, 7, or 8 bits per character are used. With 6 bits per character, 2^6 or 64 characters can be represented. With 7 bits, 2^7 or 128 characters can be represented. This allows upper- and lowercase letters and special characters to be represented. It also allows for control characters, which are used in communication. The 7-bit ASCII code is most common. The 8-bit code ($2^8 = 256$) is used for error-checking.

The ASCII code is an excellent keyboard code and finds wide application in computer and PC keyboards. The keyboard translates from the programming language (such as BASIC) into ASCII when sending a message to the computer, and from ASCII to the programming language when receiving a message from the computer. The ASCII code is given in Appendix D.

SUMMARY

The basic number systems (binary, octal, and hexadecimal) have been presented, as well as several codes (BCD, Gray, Baudot, and ASCII). These systems and codes are used for PC operations, including communication between operator and the PC, and between the PC and peripherals.

Chapter 5

Relays in Control Applications

IT WAS NOTED IN CHAPTER 1 THAT PROGRAMMABLE CONTROLLERS WERE DEVELOPED TO replace relays in control applications. In Chapter 2, the use of relays to perform the functions of logic gates was discussed. In this chapter, we examine the use of relays in control applications and introduce *relay-ladder diagrams*. Relay-ladder programming will be explored in a later chapter.

THE RELAY

The relay is nothing more than an electromagnetic switch. In a relay, an electromagnet is used to activate or deactivate the switch. This is shown in Fig. 5-1. When a current flows through the coil, it generates a magnetic field, which attracts the armature. The moving contact on the armature is drawn down to the stationary, *normally open* contact and completes the circuit. Thus, current flows to the lamp, just as if a switch in the line had been closed. When the current flow to the coil is interrupted, its magnetic field collapses, the armature is no longer attracted to the coil, and the lamp circuit is opened.

The name *relay* comes from the history of the device. The relay was first used in telegraphy. In the early days of the telegraph, dc voltage was employed. The difficulty with dc voltage was that its effective range was limited to less than twenty miles; i.e., telegraph signals were so weakened after transmission over a range of twenty miles that they were unintelligible. To avoid erecting telegraph stations every twenty miles and using human operators to relay signals, the relay was invented. Relays were placed along the telegraph lines at about twenty miles distance. They detected weak telegraph signals, and "relayed" them (with increased strength) to stations further down the line. This is illustrated in Fig. 5-2.

The use of ac voltage in telegraphy greatly increased the range between relays, of course, but relays are still used in that application. Relays are also used in a wide

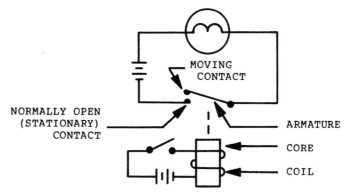

Fig. 5-1. The relay.

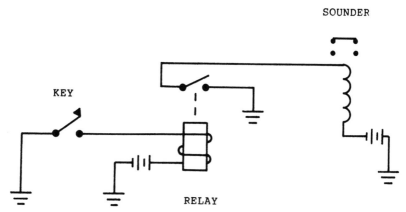

Fig. 5-2. The relay in telegraphy.

variety of other applications. The typical automobile provides an example of why a relay would be used instead of a switch. The starter of an automobile draws a large current (up to several amps) when it is turned on. A switch designed to carry that much current would be very heavy and would be difficult to throw. However, with a relay in the circuit, a very small current is needed to energize the relay coil, and the switch to the relay becomes smaller and easier to throw. The high current requirement of the starter motor is taken care of in the relay with large, high-current contacts.

The simple relay shown in Fig. 5-1 is a single-pole single-throw (SPST) relay. That is, it closes (or opens) a single set of contacts. Like switches, relays can be purchased with a variety of poles and throws. Figure 5-3 shows two examples. In Fig. 5-3(a) we have a single-pole double-throw relay. This relay has two sets of stationary contacts: those that are *normally open*, and those that are *normally closed*. "Normally" refers to the contact condition when the relay coil is not energized. Thus a normally open contact is not in contact with the moving contact when the coil is not energized. Normally open and normally closed are abbreviated N.O. and N.C., respectively. In Fig. 5-3(b), a three-

(a)

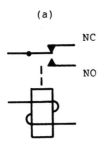

NC

NO

(b)

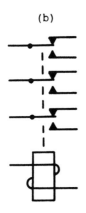

Fig. 5-3. Other relays: (a) Single-pole double-throw (SPDT) relay. (b) Three-pole double-throw relay.

pole double-throw relay is shown. This is essentially three single-pole double-throw relays, all activated by the same relay coil.

INDUSTRIAL RELAYS

Relays used in industrial control applications operate on the same general principles as the relays described above. There are, however, a few key differences. Industrial control relays are extremely rugged. They are built to survive in the sometimes-rigorous manufacturing environment. This is similar to one of the key differences between the personal computer and the programmable controller. Also, industrial control relays almost always provide both normally open and normally closed contacts; i.e., they are almost always double-throw relays. Finally, the design of the armature and moving contacts is different from those relays just described. (See Fig. 5-4.)

This figure shows that the armature moves the moving contacts up or down. The moving contacts are now paired; i.e., there are two of them. This helps to dissipate the transient voltages (generated by the arc at the contact point) more quickly than with a single moving contact.

RELAY CONTROL EXAMPLES

Relays exist in one of two states: on or off. Thus relays are on-off controllers. This section presents two examples of how relays are used in on-off control applications.

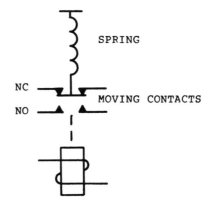

Fig. 5-4. Industrial relay construction.

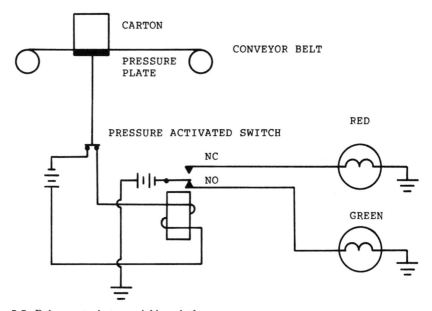

Fig. 5-5. Relay control at a weighing platform.

The first example is very simple and involves a weighing platform at a shipping station (Fig. 5-5). As filled, sealed cartons proceed down a conveyor belt, they pass over a pressure plate. If the carton is completely filled, the pressure plate activates a switch that energizes a SPDT relay. The normally open contacts of the relay are connected to a green lamp, while the normally closed contacts are connected to a red lamp. Thus, if the box is filled properly, the green light appears and the shipping clerk knows that the box is ready to be shipped. If the red light stays illuminated, the shipping clerk knows that the box is not properly filled. The box is then set aside to be refilled before shipping.

Note that although the refinements shown in Fig. 5-4 are not included in the relay shown in Fig. 5-5, it is assumed that the relay is an industrial control relay. This assumption will be made throughout the rest of this book.

The second example involves the control of steam temperature. Steam temperature is controlled with water. To lower the temperature of steam, one simply injects water into the steam line. Many chemical reactions are extremely sensitive to temperature. If reaction temperature (i.e., steam temperature) is too high, many reactions become explosive. Thus preventing excessive high temperatures is an important control function.

Figure 5-6 shows a very simple process for the control of steam temperature. Note that a single-pole single-throw relay is used. Two objects are placed in the steam line. The first is a feedwater line. When the feedwater pump is turned on, cooling water is sprayed into the steam line. The second object in the steam line is a *thermocouple*. A thermocouple is a device that produces a dc voltage that is proportional to temperature.

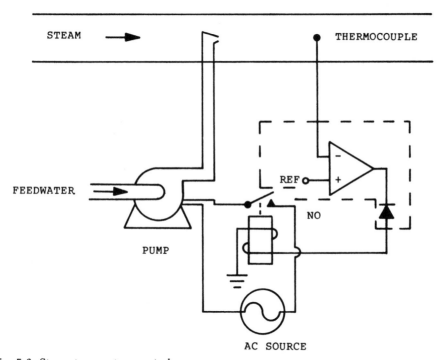

Fig. 5-6. Steam temperature control.

The control system operates as follows. The thermocouple produces a positive dc voltage that is proportional to temperature. Its signal is fed into the inverting input of an operational amplifier. A positive reference voltage, which is equal to the temperature at which the relay should be activated, is fed into the noninverting input of the operational amplifier. The operational amplifier in this mode acts as a *comparator*. If the reference voltage (critical temperature) is greater than the thermocouple voltage (actual temperature), the output from the comparator will be positive, which reverse-biases the diode. No current will flow to the relay. If, however, the thermocouple voltage exceeds the reference voltage (i.e., the critical temperature is exceeded), then the output of the comparator goes negative, which forward-biases the diode and triggers the relay.

When the relay is triggered, an ac voltage is applied to a constant-speed pump that sprays cooling water into the steam line. This will reduce the temperature of the steam and the voltage from the thermocouple. When the thermocouple voltage falls below the reference voltage (i.e., the actual temperature falls below the critical temperature), the relay will be deenergized, and the pump will turn off.

The portion of Fig. 5-6 that is enclosed by dashed lines can be thought of as a *temperature-activated switch* that is normally open. At a certain temperature, the switch closes, allowing current flow to the coil of the relay.

Note that the relay-control scheme described above offers only on-off control. It does not allow incremental control of steam temperature.

In the next section, a convenient shorthand for the description of relay-control systems (such as the two examples given above) will be presented.

LADDER DIAGRAMS

The two relay-control examples presented in the previous section can be represented by a convenient shorthand notation. This notation is the *ladder diagram*. The two following examples should clarify the concept of the ladder diagram.

Figure 5-7 shows the *relay ladder diagram* for the control circuit in Fig. 5-5. The term *ladder* derives from the appearance of the diagram. The vertical lines in the diagram represent power lines; in this case, the left line is at a positive dc potential and the right line is negative. In many instances, ac power lines are used. Power flows through the "rungs" of the ladder when the circuit is completed. For example, in rung 1, when the pressure-activated switch PS is closed, power flows through the coil R1 of the relay.

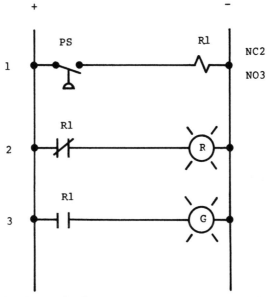

Fig. 5-7. Relay ladder diagram for Fig. 5-5.

To the right of the negative power line at rung 1 are the cross-referencing comments "NC 2" and "NO 3." These comments indicate the rung numbers that represent the contacts of relay R1. In this case, rung 2 describes the normally closed contacts of the relay, while rung 3 describes the normally open contacts of the relay.

Rung 2 in Fig. 5-7 shows the symbol for a normally closed contact and a red lamp, while rung 3 shows the symbol for a normally open contact and a green lamp. The "R1" above both sets of contacts simply reminds us that these are contacts for relay number 1, whose coil circuit is shown in rung 1.

The three rungs in Fig. 5-7 completely describe the control system of Fig. 5-5. When PS is closed, the relay coil is energized and power flows through rung 3, illuminating the green lamp. When PS is open, the relay coil is not energized, and power flows through rung 2, illuminating the red lamp.

Figure 5-8 is a second example of a relay ladder diagram. This figure describes the control circuit shown in Fig. 5-6, a circuit that controls steam temperature. The circuitry enclosed in dashed lines in Fig. 5-6 is represented in Fig. 5-8 as a normally open temperature-activated switch (TAS) in rung 1 of the ladder diagram. When the temperature-activated switch is closed (i.e., when the steam temperature exceeds a critical temperature), the coil of the relay R1 is energized.

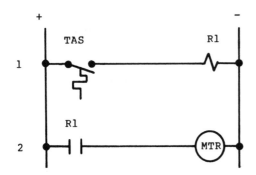

Fig. 5-8. Relay ladder diagram for Fig. 5-6.

Since the relay is a SPST relay, only one additional rung is needed to complete the relay ladder diagram. This appears as rung 2 in Fig. 5-8. When the relay coil is energized, the normally open contacts of the relay are closed, and power is supplied to the pump motor (MTR in Fig. 5-8). Thus cooling water is supplied to the steam line until the TAS in rung 1 opens again.

While the relay ladder diagrams are descriptive, they are often replaced with *contact ladder diagrams*. Contact ladder diagrams show only contacts and outputs (which are indicated by parentheses). Contact ladder diagrams for Figs. 5-7 and 5-8 are shown in Figs. 5-9 and 5-10, respectively.

SUMMARY

This chapter has presented two examples of relays used in control applications. Relay and contact ladder diagrams have been introduced. A later chapter will further develop relay and contact ladder diagrams and their use in programming PCs.

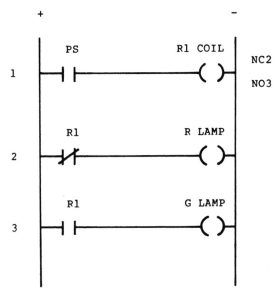

Fig. 5-9. Contact ladder diagram for Fig. 5-7.

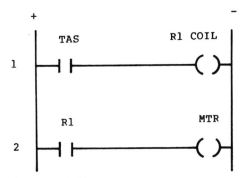

Fig. 5-10. Contact ladder diagram for Fig. 5-8.

Chapter 6

The Central Processing Unit

THE PRECEDING FIVE CHAPTERS PRESENTED BASIC INFORMATION NECESSARY FOR AN understanding of programmable controllers and their operation. In this chapter, and the two that follow, we begin an examination of the hardware associated with programmable controllers. This chapter will examine the central processing unit. Chapter 7 will examine input/output configurations, and Chapter 8 will explore peripherals.

The *central processing unit* (CPU) is composed of three elements: the power supply, the microprocessor (or processor), and memory. We will examine each of these elements in turn.

POWER SUPPLIES

The power supply is one of the most important elements of the central processing unit, because without it, processor and memory circuits could not function. It is also the simplest element of the CPU to select. Unfortunately, it is one of the most overlooked elements.

The power supply, in addition to providing power to the processor and memory, is often used to provide power to input/output (I/O) modules and for communications between the processor and remote I/O modules. Because of this, it is often necessary to select a power supply that is capable of delivering more power than is required for operation of just the processor and memory.

In most cases, PC power supplies require ac input voltages. There are some, however, that use dc input voltages. These are more commonly used in applications in which ac voltage is not available, such as at on-site field operations. Generally, ac-input power supplies accept 120 or 220 volts ac input, while dc power supplies accept 125 or 24 volts dc input.

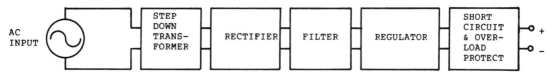

Fig. 6-1. Block diagram of a PC power supply.

A block diagram of a typical PC power supply is shown in Fig. 6-1. The ac line voltage is stepped down by the transformer and rectified (converted to dc voltage) in the rectifier section. After the rectifier section, the voltage is filtered and *regulated*. The regulator is an electronic device that maintains the output voltage at a constant level regardless of the load on the power supply. After the regulator is the short circuit and overload protection circuitry. This prevents a short circuit or an overload from destroying the power supply (or the microprocessor or memory circuits).

Most of the elements found in the power supply shown in Fig. 6-1 are found in normal power supplies. However, there are certain differences between a normal power supply and a PC power supply. PC power supplies are used in industrial plants where ac line voltage experiences fluctuations, both in voltage level and frequency. These fluctuations are caused by the starting up or shutting down of large pieces of equipment (motors, pumps, etc.). A normal power supply would not function properly under such conditions of fluctuation in voltage and frequency; however, a PC power supply is designed to operate under these conditions. Additionally, PC power supplies are designed to shut the microprocessor and memory down if voltage variations or frequency variations become excessive.

One additional difference between a normal power supply and a PC power supply is that the PC power supply is designed to operate in areas of high *electromagnetic interference (EMI)*. Electromagnetic interference is often generated by heavy equipment operating near the PC. PC power supplies are usually protected from EMI by *isolation transformers*. The isolation transformer is installed between the ac power line and the programmable controller.

As with standard power supplies, the power supply should be *sized* according to the *load* that will be connected to the power supply. Many programmable controllers include the power supply as an integral part of the unit. With these PCs, there will be no need to select a power supply. For PCs in which the power supply is not included, it is important to select a power supply that will handle the maximum load of the microprocessor and memory, plus some of the load from the input/output modules. For PCs that use many I/O modules, *auxiliary,* or *remote* power supplies will be needed. These supplies should be selected in accordance with the load requirements of the I/O modules.

The easiest way to determine the proper size of the power supply is through *current summation*. Simply sum the current requirements for all I/O modules to be used, and select a power supply that is capable of supplying the total current or slightly more than the total current.

Often PC power supplies include *battery backup systems*. This is particularly useful for certain types of memories, called *volatile memories*, which do not maintain programs when the power supply is interrupted. Battery backups employ many different kinds of batteries (e.g., carbon-zinc, alkaline, nickel-cadmium, lithium, lead-acid, etc.). Generally, carbon-zinc or alkaline batteries are used in the majority of PCs, although lithium batteries are quickly becoming popular. Carbon-zinc, alkaline, and lithium batteries are called *primary* batteries because they are not rechargeable. The nickel-cadmium and lead-acid batteries are called *secondary* batteries because they can be recharged.

The use of batteries in PC systems requires that proper maintenance and monitoring be performed. Batteries that become discharged often leak electrolyte, and this can corrode the contacts and wiring in the power supply.

THE MICROPROCESSOR

At the heart of the microprocessor (or processor) section of the PC is the integrated-circuit microprocessor chip. The microprocessor chips used in programmable controllers are exactly the same as those used in computers. These are generally the Z80, 8080, 8086, 6800, or 9900 family of microprocessors. If you understand how a microprocessor functions in a personal computer, then you understand how a microprocessor functions in a programmable controller. (The arithmetical and logic operations of microprocessors were demonstrated in Chapter 4.) The differences between the PC and the personal computer are in the power supply, noise protection circuitry, programming language, and nature of the input/output circuitry.

The microprocessor forms the intelligence of the programmable controller. It is, as previously noted, an integrated circuit that performs the mathematical operations, handles the data, and executes the diagnostic routines associated with programmable controllers. In short, the microprocessor controls all activities of the PC.

The microprocessor is controlled by a system of programs known as the *executive*. Executive programs are stored permanently in the PC and supervise the operation of the PC. This supervisory function includes control, data processing, and communication with I/O modules and peripherals.

In addition to being controlled by the executive program, the microprocessor is also controlled by *application programs,* or *user programs*. These programs are the actual instructions for a particular control application. Usually these programs are in the form of ladder-logic diagrams. They can also be written in a higher-level language, such as the languages that will be described in a later chapter.

Some programmable controllers use an approach known as *multiprocessing*. This is simply a method of reducing the time required to implement a program. With multiprocessing, several microprocessors are used to perform control tasks. An example of this is the use of an *intelligent I/O module*, such as the *proportional-integral-derivative (PID)* control module. This module is programmed with the equations for proportional-integral-derivative control and is generally located away from the CPU. The primary advantage of multiprocessing is that the time necessary to process instructions and control functions is greatly reduced.

Microprocessors generally are categorized by *word size*. As was discussed in Chapter 4, a word is a group of one or more bytes. Typically, a byte is a group of 8 bits; however, words can be composed of 4, 8, or 16 bits. Word length is an important characteristic of PCs: a longer word length allows faster manipulation of data, since more data are handled in one operation with a longer word.

The microprocessor is responsible for accepting inputs from the I/O modules, manipulating the data included in the inputs, and updating outputs. This process is known as performing a *scan*. PCs are also categorized by *scan time*. Typically, scan time (the length of time required to perform a scan) ranges between 1 and 100 milliseconds (msec).

It is important to consider the scan time of a PC when selecting the proper PC for a particular operation. If input signals change quicker than the scan time, the PC will not be able to act upon the input data. Thus it is important that a PC be selected with the appropriate scan time (i.e., a scan time that is faster than changes in input data).

Microprocessors also include *diagnostic programs*. These programs detect failure in communications, system operation, etc. A failure in any one of these areas would be detected by the diagnostics in the microprocessor and would activate an alarm circuit to signal a failure.

MEMORY

Programmable controllers, by definition, use a programmable memory for storage of instructions that are used to implement specific functions. Thus the memory is an important part of the central processing unit. This section will discuss the types of memory available for use with PCs and the *architecture* of memories used in PCs.

The similarity between PCs and personal computers continues within the memory section. The types of memories used in PCs are the same as those used in computers.

Memory Systems

Generally, memory systems fall into one of two categories: *volatile* or *nonvolatile*. Volatile memory is memory that loses its contents when the power supply is interrupted. Generally, volatile memories are easily altered or erased. Nonvolatile memories retain their contents even if the power supply is interrupted. Nonvolatile memories usually are difficult, if not impossible, to alter. It is not unusual to find both volatile and nonvolatile memories in programmable controllers.

Random-access memory (RAM) is a volatile semiconductor memory. Sometimes RAM is categorized by the semiconductor technology used in its construction; for example, *CMOS RAM* (for Complimentary-Metal-Oxide Semiconductor Random-Access Memory). RAM, as a volatile memory, loses its contents when the power supply is interrupted. The contents stored in RAM are easily changed (i.e., RAM is easy to reprogram). While random-access memory is not used to store the executive programs, it does work well for the storage of input data and application programs. RAM is a relatively fast memory.

There are two types of random-access memory that are nonvolatile. These are *core memory* and *NOVRAM*. Core memory, shown in Fig. 6-2, consists of small magnetic donuts that are magnetized in either a clockwise or counterclockwise direction, depending upon the flow of current in the wires that pass through the center of the donuts. Each donut represents one bit of memory. Magnetization in the clockwise direction represents a 1, while magnetization in the counterclockwise direction represents a 0. Since the donuts maintain their magnetization for long periods of time, core memory is nonvolatile. Core memory is an old technology, and is slower, more expensive, and less compact than modern semiconductor RAM.

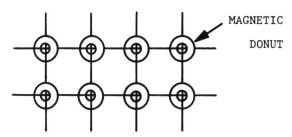

Fig. 6-2. Core memory showing two 4-bit words.

NOVRAM is an acronym for NOnVolatile Random-Access Memory. It employs conventional semiconductor RAM and a nonvolatile semiconductor memory known as *EEPROM*. (EEPROM will be discussed later.) Both memories are fabricated on a single chip. NOVRAM does not find much use in programmable controllers today. When it is used, it is generally in small programmable controllers.

All nonvolatile memories (apart from core memory) represent some form of *read-only memory (ROM)*. Generally, read-only memory is not easily reprogrammed. Once a program is entered into ROM, it stays there permanently, regardless of the status of power supply to the ROM chip.

There are several types of read-only memory. These are: PROM, EPROM, UV-EPROM, EEPROM, and EAROM.

PROM is an acronym for Programmable Read-Only Memory. PROM cannot be altered once it has been programmed (i.e., once data or instructions have been stored in the memory). In order to change the contents of a PROM memory, the entire PROM chip must be replaced with one that contains the desired data. *EPROM* is an acronym for Erasable Programmable Read-Only Memory. An EPROM chip can be erased in one of two ways. It can be erased by exposure to ultraviolet light, in which case it is called UV-EPROM; or it can be erased with an electric charge, in which case it is called electrically erasable programmable read-only memory, or EEPROM.

The erasable and reprogrammable nature of the two EPROMs makes them popular memories. Reprogramming an EPROM is not as easy as reprogramming a RAM, however, since the EPROM must be *totally* erased before reprogramming can occur. Also, EPROMs have finite lifetimes; i.e., they can be erased and reprogrammed a limited number of times. Nevertheless, their use in programmable controller memories has increased.

A third type of ROM is the EAROM, or Electrically Alterable Read-Only Memory. Application of a relatively low voltage to a pin on the EAROM chip results in erasure. Thus, EAROMs can be quickly erased without having to be removed from the circuit board. EAROMs are used in very few programmable controllers today. It has the disadvantage of being a relatively slow memory.

Memory Architecture

The architecture of a PC's memory specifies how the various memory systems described above are organized and used by the PC in performing control functions. Memory architecture is usually shown schematically by a *memory utilization map*, such as the one shown in Fig. 6-3. A glance at the memory map indicates that PC memory is divided into three major areas: system memory, input/output status memory, and application memory.

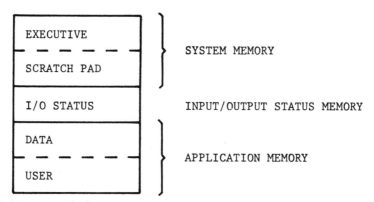

Fig. 6-3. Memory utilization map.

System memory can be subdivided into two types of memory: *executive memory* and *scratch-pad memory*. Executive memory contains the executive program (or executive operating system). Since the executive is provided by the PC manufacturer and rarely changes, executive memory is ROM (actually, PROM). It was noted earlier that the executive program controls the operation of the PC. It provides the translation between the high-level programming language (such as BASIC) and the binary machine language, scans the PC to update system status, and reads inputs and updates outputs. In administering its functions, the executive often needs an area of memory for the temporary storage of data. The portion of memory reserved for this purpose is called scratch-pad memory. It is composed of RAM and is reserved for the exclusive use of the executive (i.e., it is not accessible to the PC user).

Input/output status memory is accurately described by its name. It is a portion of RAM that is set aside for storage of current input/output statuses. Since it is the executive program that requires updating of I/O statuses, the I/O status memory can be thought of as a part of the system memory.

Application memory is the final area of PC memory. Like the system memory, application memory can be subdivided into two types: *data memory* and *user memory*. These memories hold the data used by the microprocessor in fulfilling its control functions and the user program (or instructions) that direct the microprocessor to perform its control functions.

Data memory generally holds preset values. For example, the user program may contain a statement that sets a counter or timer to a particular value, based upon I/O statuses. The program may also instruct the microprocessor to perform certain math functions (addition, subtraction, etc.) on data that are obtained. The data memory then would be used to store the preset values for timers and counters, and the instructions for data manipulation (math functions). Since the values stored in data memory need to be changed frequently, data memory is RAM.

User memory is the memory that is most accessible to the user or programmer. The user program provides the specific instructions for control of processes and provides the "programmable" feature of programmable controllers. User memory is scanned by the microprocessor for specific instructions when the microprocessor is directed to scan by the executive. This scan of, or communication with, user memory is accomplished through the use of two *buses*. (A bus is a group of lines used for data transmission or power distribution.) The first bus addresses a particular location in memory and is known as the *address bus*. The second bus is used to transmit the data stored in the memory and is called the *data bus*. Since the user program may need to be changed frequently, user memory is RAM.

Although RAM is the predominant memory type for user memory, EPROM and EAROM are sometimes used. Both EPROM and EAROM are not as easily reprogrammed as RAM; therefore, EPROM and EAROM are used in PCs whose user programs seldom, if ever, change.

General Comments on Memory

The basic structural units of memory - the bit, byte, and word - were discussed previously. While microprocessors are generally characterized by word size, memory is characterized by number of words that can be stored. The amount of memory is designated by the letter K. One K of memory provides the capacity for the storage of 1024 words. Likewise, 8K of memory provides the capacity for the storage of 8192 words. The words, of course, may vary in length. Typically, word length is either 8-bit or 16-bit. Word length will, of course, affect the total memory capacity of the PC. A 1K memory of 8-bit word length can store only one half the information stored in a 1K memory of 16-bit word length.

Memory is an important consideration in the specification of a microprocessor. For example, the scan time of a microprocessor (discussed previously) is generally specified in terms of time per K of programmed memory.

When memory is specified for a PC, it is important that the distribution of the memory be understood. Although many PC suppliers specify only application memory, others may include system memory and I/O status memory in the total amount of memory specified. The supplier's literature should be consulted in order to determine the category of memory specified.

SUMMARY

The three basic components of the central processing unit have been discussed. These components are very similar to the components found in a personal computer. Certain differences exist. Power supplies for programmable controllers are generally more rugged and are more protected than power supplies found in personal computers. PC microprocessors are protected from noise and other interferences. Memory requirements for PCs are somewhat different from memory requirements for personal computers, especially as regards input/output status memory.

Chapter 7

Input/Output Interfaces

LET US ASSUME THAT WE WANT TO USE A PC TO CONTROL STEAM TEMPERATURE. A RELAY control circuit for that purpose was given in Fig. 5-6. As a first attempt, we use the circuit of Fig. 5-6, but replace the temperature-activated switch (comparator) and relay with a PC. The result is shown in Fig. 7-1. After several hours we notice that the pump motor has never started. We measure the steam temperature in the line and discover that it is dangerously high. We conclude that the PC circuit does not work, and then we manually turn on the pump motor.

Why did the PC circuit in Fig. 7-1 not perform as expected? There are two very good reasons. First, the output of a thermocouple is typically on the order of millivolts. Since the PC in Fig. 7-1 uses 0 volts dc as a 0 level and +5 volts dc as a 1 level, the output of the thermocouple is insufficient to provide a 1 input to the PC. Second, the PC is incapable of driving the pump motor because (1) it does not provide enough output power, and (2) it cannot be connected directly to an ac source.

In order to make the circuit in Fig. 7-1 perform properly, the input signal to the PC must be treated to boost it to a level that is compatible with the CPU's logic circuitry, and the output signal from the CPU must be isolated from the ac source and boosted to an appropriate power level. These functions are performed by *input/output interfaces*, as shown in Fig. 7-2. The I/O interfaces thus provide the means by which the PC interacts with input and output devices.

This chapter will explore the various I/O interfaces that are used with PCs. Generally, these interfaces fall into one of the following classes: discrete, numerical data, analog, and special.

DISCRETE INTERFACES

The discrete interface is the oldest and most commonly used class of I/O interfaces. The discrete interface is used with input/output devices that provide or require discrete

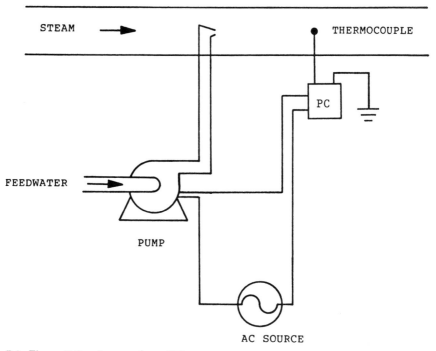

Fig. 7-1. Figure 5-6 redrawn using a PC.

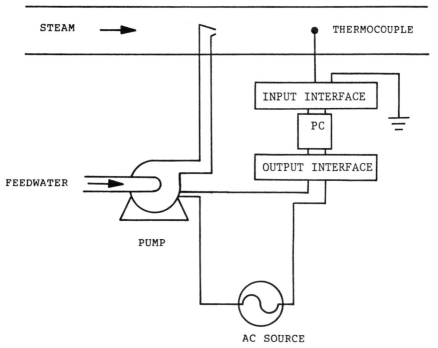

Fig. 7-2. Control of steam temperature with a PC and I/O interfaces.

signals. An example of a discrete input device is a switch that is either on or off. Similarly, a discrete output interface is used to connect the PC to an output device that can be in only one of two states (i.e., on or off), such as lights, relays, or alarms.

One of the primary functions of an input interface is to match voltage levels between an input device and the PC. A block diagram of an input interface that performs this function is shown in Fig. 7-3. The interface in this figure accepts an ac input voltage signal and converts it to a dc level suitable for the PC's microprocessor. For example, the input may be from a device that provides a 120-volt ac signal. The first stage of the interface rectifies and filters the signal. The second stage of the interface is the signal threshold detector. If the ac signal is greater than the background noise level (i.e., if the signal is sufficient to indicate that the input sensor is functioning in the on mode), the signal is treated as a 1 signal and an on voltage at a level appropriate for the PC's microprocessor is generated. The final stage of the interface is the isolation stage, which may be a transformer or optical coupler. The isolation stage removes any electrical connection between the input device and the PC. This prevents damage to the PC that may occur from anomalies such as voltage spikes.

A discrete output interface is shown in Fig. 7-4. As can be seen from the figure, the output interface is similar in construction to the input interface. The signal from the PC is electrically isolated from the output device. The signal enters a switching section that converts the PC's dc signal to an ac voltage at a level suitable for use by the output device.

Although Figs. 7-3 and 7-4 show input/output interfaces that convert ac to dc and dc to ac, respectively, not all discrete I/O interfaces are used to convert between ac and dc. In many cases, the input and output devices connected to the PC use dc voltage; however, the voltage level may not be the same as that required for the operation of the PC. Additionally, current inputs and outputs might be required. In these instances, the I/O interfaces would be used to convert the dc voltages involved to appropriate levels,

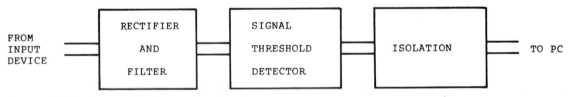

Fig. 7-3. Block diagram of a discrete input interface.

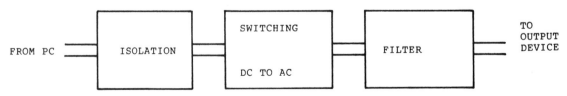

Fig. 7-4. Block diagram of a discrete output interface.

or to convert current inputs to voltages for use by the PC, and voltage outputs from the PC to currents for use by the output devices. Other I/O interfaces are used to handle TTL (*t*ransistor-*t*ransistor *l*ogic) voltages (5 volts dc).

NUMERICAL DATA INTERFACES

Discrete interfaces handle I/O data that consist of single bits: either 1 or 0. Numerical data I/O interfaces handle inputs and outputs that consist of many bits. In other words, numerical data interfaces are capable of handling data in the form of multiple bits, such as binary-coded decimal inputs and outputs.

A typical multibit input would be data from a thumbwheel switch. A BCD thumbwheel switch that is set to the number 9, for example, would generate the BCD number 1001. These four bits would be processed by the input interface and transferred to the PC in a format that is acceptable to the PC.

An example of a multibit output device would be a seven-segment display, which requires four bits of data to light the display.

In summary, the numerical data I/O interface is very similar to the discrete I/O interface in that it handles discrete bits of data. The difference, of course, is that the numerical data I/O interface handles multiple bits, whereas the discrete I/O interface handles only single bits.

ANALOG INTERFACES

Many input devices, such as thermocouples and pressure transducers, provide a continuously varying analog output signal. In order for these signals to be processed by the PC, the analog signals must be converted to digital signals. Also, once the PC has operated on the information received from the input sensor, a digital output signal will be provided. This digital PC output signal may not be suitable for the control of certain output devices, such as motors and chart recorders. Thus, in certain cases, the PC's digital output signal will need to be converted to an analog signal. I/O interfaces that convert an analog signal to digital or a digital signal to analog are called analog interfaces. These are simply digital-to-analog converters (DAC) or analog-to-digital converters (ADC).

A typical analog input interface is the thermocouple interface. Thermocouples usually generate output signals on the order of millivolts. An analog input interface for a thermocouple would thus be capable of accepting a very low-level input signal, filtering the signal, amplifying the signal to a level appropriate for the PC, and converting the analog signal to digital with an ADC. If the output signal from the PC is to be used to control an electric motor, the PC's output signal would be converted to analog (with a DAC) and boosted to the level necessary to drive the electric motor by the analog output interface.

As these examples point out, analog I/O interfaces are essentially DACs or ADCs that include amplification and filtering capabilities.

SPECIAL INTERFACES

The discrete, numerical data, and analog interfaces will satisfy the majority of the PC's data processing requirements for input and output signals. There are a few special interfaces, however, that deserve mention. These interfaces are sometimes called *intelligent interfaces* in that they perform certain tasks independent of the PC. The use of intelligent interfaces in process control is often known as *distributed processing*. A few of these special interfaces will be discussed.

Data-processing interfaces are modules that perform some of the functions of a microprocessor. A data-processing interface can be thought of as a small, dedicated computer whose job is to perform preset calculations on data received from sensors or other devices connected to the interface. The interface would be capable of storing data and would perform mathematical calculations (e.g., addition or subtraction of input signals, square roots of input signals, etc.) on the data. The use of data processing interfaces eliminates some of the demands on the PC's microprocessor, thus freeing the PC so that it can handle larger control functions.

As will be seen in the next chapter, peripherals constitute an important part of the PC system. The use of peripherals (such as printers, video display terminals, etc.) requires special interfaces. A typical interface for peripherals would be an *ASCII interface*, which allows the transmission of alphanumeric data between the PC and peripherals. The ASCII interface is not an intelligent interface, in that it does not perform microprocessor or control functions. It does, however, speed up communications between peripherals and the PC.

One of the most popular of the special I/O interfaces is the *proportional-integral-derivative (PID) interface*. The PID interface is used to provide PID control for various processes that require continuous, closed-loop feedback control. Since PID control is a very important concept, it will be discussed in detail in the following section.

Proportional-Integral-Derivative Control

PID control can best be described with an example. Figure 7-5 shows a system in which temperature must be controlled. A reactant stream is fed to a reactor, which converts the reactant into a product. If the temperature of the reactant stream is less than a specified temperature T, the yield of product produced by the reactor is diminished, so that the reactor becomes uneconomical to operate. At a temperature greater than T, the reaction becomes quite vigorous, eventually approaching an explosive level. Thus, economic and safety considerations require that the reactant stream be maintained at or near temperature T.

The simplest method of controlling reactant stream temperature T is shown in Fig. 7-6. In this figure, a set-point temperature T_s is fed to a proportional controller, as is the actual temperature T of the reactant stream. (The proportional controller can be an intelligent interface or a PC programmed for proportional control.) The proportional

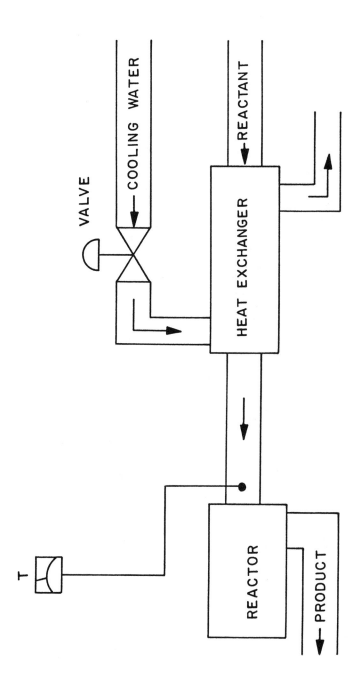

Fig. 7-5. The model system.

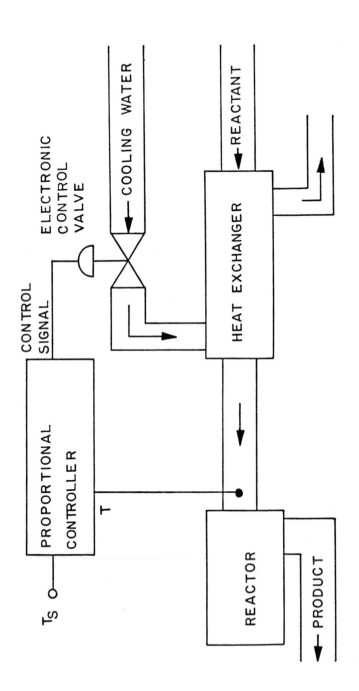

Fig. 7-6. The proportional control scheme.

controller measures the error ϵ in the temperature, where $\epsilon = T_s - T$. The output of the proportional controller is given by Equation 7-1:

$$e = K \epsilon + e_c \qquad (7\text{-}1)$$

where

e	$=$	voltage output from the proportional controller (i.e., the control signal)
K	$=$	gain
ϵ	$=$	error signal $= T_s - T$
e_c	$=$	constant
T_s	$=$	set-point temperature (i.e., desired temperature)
T	$=$	measured temperature

From Equation 7-1 you can see that the control signal e is proportional to the error signal ϵ. In fact, Equation 7-1 is just the equation for a straight line. Thus the control signal is linearly proportional to the error signal.

The temperature-versus-time plot for the proportionally controlled system in Fig. 7-6 is shown in Fig. 7-7, and the plot is compared with that of an uncontrolled system. In this figure it is assumed that the proportional control system is *ideal*; i.e., that Equation 7-1 is obeyed exactly. It can be seen from the figure that, with no control, the temperature of the reactant stream increases constantly, eventually approaching the explosive limit. With proportional control, however, the temperature oscillates back and forth about the set-point temperature, eventually reaching equilibrium at T_s.

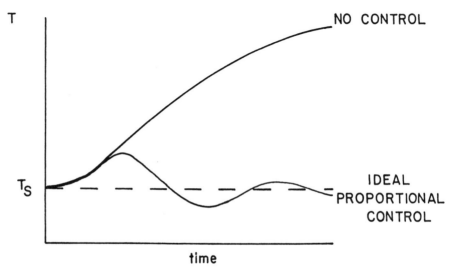

Fig. 7-7. Comparison of ideal proportional control with no control.

The oscillatory approach to the set-point temperature is caused by time lags in the system. The control valve that allows cooling water to flow to the heat exchanger does not open instantaneously, and the heat exchanger does not drop the reactant stream temperature immediately. Because of the time lags in the valve and heat exchanger operations, the controller overcorrects for temperature; i.e., the controller drives the reactant stream temperature below the set-point value. The controller then senses the new error and attempts to compensate for it. Again, the controller overshoots the mark, but not by quite as much as before. This process continues until the reactant stream temperature is very nearly equal to the set-point temperature. Thus the action of the controller as shown in the figure appears to be that of a damped oscillator.

In Fig. 7-8, the response of a real (i.e., nonideal) proportional controller is shown. As can be seen from the figure, the temperature of the reactant stream reaches a steady state value higher than that of the set-point temperature. The difference between the steady state and set-point temperatures is called the *offset*.

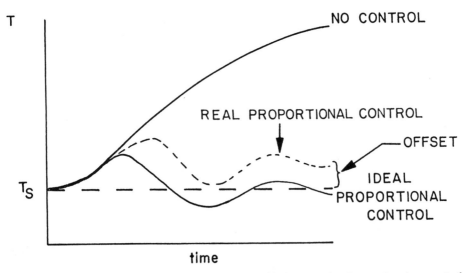

Fig. 7-8. Comparison of real proportional control with ideal proportional control and no control.

The offset occurs because Equation 7-1 is not followed exactly. Equation 7-1 is the steady state control equation. It assumes that the change in temperature from the set-point value occurs instantaneously. This is not usually the case. The effect of the finite time required to make a change in temperature is the addition of a term to Equation 7-1, as shown in Equation 7-2:

$$e = K \epsilon + e_c + f(t,K) \tag{7-2}$$

where the function $f(t,K)$ is called the *transient response*. The transient response is a function of time and gain and is largely responsible for the offset.

In the model system, an offset cannot be tolerated. A better control scheme, one that removes the offset, is the *proportional-integral* control scheme. In proportional-integral control, the proportional controller in Fig. 7-6 is replaced with a proportional-integral controller, which adds a term to Equation 7-1 that corrects for transient response:

$$e = K\epsilon + K/\tau \int_{t=0}^{t} \epsilon\, dt + e_c \tag{7-3}$$

where τ is a time delay.

The control signal e is now proportional to the error and to the time integral of the error. If it is assumed that a unit change in temperature occurs (i.e., that $\epsilon = 1$), the response of the controller becomes

$$e = K + (K/\tau)t + c \tag{7-4}$$

where c is a constant that is determined by initial conditions. Thus with a unit change in temperature, e changes suddenly by an amount equal to K (proportional action) and then changes linearly with time at a rate equal to K/τ. In this way, the offset due to transient response is removed.

Figure 7-9 shows a comparison of no control, proportional control, and proportional-integral control. It can be seen that the offset in temperature is removed by proportional-integral control and that the steady-state value approaches the set-point value. Proportional-integral control approximates ideal proportional control, with the same degree of oscillatory behavior. Although the offset is removed, it takes a long time for the oscillations about the set point to dampen.

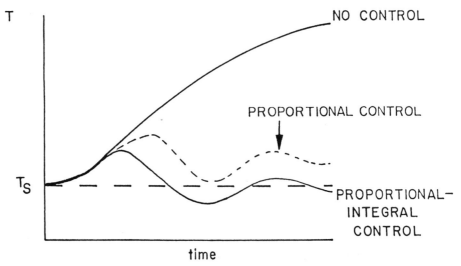

Fig. 7-9. Comparison of no control, proportional control, and proportional-integral control.

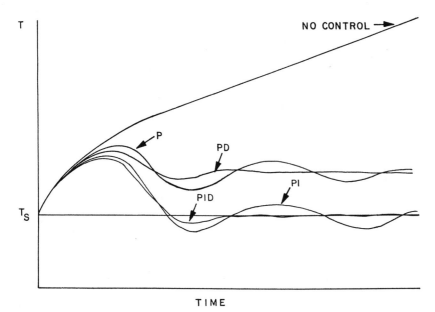

Fig. 7-10. Comparison of the proportional-integral-derivative control scheme with all others.

The oscillatory behavior can be removed by the inclusion of a term that is proportional to the derivative of the error. When this term is included, the control equation becomes:

$$e = K\epsilon + K/\tau \int_{t=0}^{t} \epsilon\, dt + K\tau\, d\epsilon/dt \tag{7-5}$$

Control signal e is now proportional to the error signal, to the integral of the error signal, and to the derivative of the error signal. The first term in Equation 7-5 corrects for the difference between the actual and set-point temperatures, the second term eliminates the offset, and the third term anticipates the change in T and dampens out the oscillations. This is the PID control scheme.

A comparison of no control, proportional control, proportional-integral control, proportional-derivative control, and proportional-integral-derivative control is shown in Fig. 7-10.

A PID interface implements Equation 7-5 to control various processes. PID control can also be performed by a PC programmed with the PID control algorithm; however, the use of a PID interface relieves the PC of one of its functions. This allows the PC to be utilized in controlling other process steps.

SUMMARY

Without input/output interfaces, the PC would be unable to perform its control functions. The basic types of interfaces have been described. These include the discrete, numerical data, analog, and PID interfaces.

Chapter 8

Peripherals

THROUGHOUT THIS BOOK I HAVE DRAWN COMPARISONS BETWEEN PROGRAMMABLE CON-
trollers (PCs) and personal computers. Such a comparison is particularly useful in
highlighting the functions of *peripherals*, or external devices that are connected to the
PC. A very basic personal computer system consists of four components: the computer
itself (CPU); a keyboard for data and program entry; a video-display terminal (VDT)
to display the program or system status; and a line printer for hard copies of programs
or data. In this simple system, the keyboard, VDT, and printer are all personal computer
peripherals.

PCs also use peripherals that perform functions similar to those described above.
There are three basic classes of PC peripherals: programming and data entry devices;
display devices; and documentation devices. This chapter will discuss each class in turn.
Additionally, the important topic of communications between peripherals and the PC will
be discussed.

PROGRAMMING AND DATA ENTRY DEVICES

As the name implies, programming and data entry devices allow the PC operator
to enter programs, preset values, and the like into the PC's memory. There are many
of these devices on the market today.

CRTs, or *cathode ray tubes*, can be thought of as a combination of the video display
terminal and the keyboard of the personal computer system described previously. The
CRT contains all the electronic circuitry necessary for communication with the PC. CRTs
are extremely useful: the display can present large amounts of information (limited only
by the size of the screen), and the presence of a complete keyboard allows easy access
to the PC's memory. Also, while some CRTs are stationary, others are portable.

There are two basic types of CRTs available: dumb and intelligent. Of the two types, the dumb CRTs are the older, the less expensive, and the more flexible (in terms of compatibility with a wide variety of different PCs). Dumb CRTs do not contain their own microprocessor or memory. They must be connected with the PC before they can operate properly. They are used as terminals and allow *on-line programming* only. Intelligent CRTs, on the other hand, contain their own microprocessors and memories. Use of an intelligent CRT would allow the programmer to write and edit programs without having to be connected to the PC. This is *off-line programming*. Intelligent CRTs are much more expensive than dumb CRTs and are usually not compatible with more than one brand of PC.

Personal computers can be used as data entry and programming devices. They function essentially as intelligent CRTs.

For some small PCs, *miniprogrammers* offer an inexpensive, portable alternative to the CRT. These are hand-held devices that resemble pocket calculators, except that their displays are somewhat larger and the keyboards contain special programming keys. Miniprogrammers can be either dumb or intelligent. They are used to enter programs into the PC's memory and for program editing. Generally, miniprogrammers are compatible only within a manufacturer's family of PCs.

Program loaders are devices used to load or reload a program into the PC's memory. Program loaders are not programming devices; i.e., they are not used for writing or editing programs. They merely provide a convenient method for loading a program into the PC's memory. One of the oldest and most popular of the program loaders is the cassette tape recorder. A written and edited program is fed into the cassette tape recorder, which is then transported to the PC site, where its contents are unloaded. Another program loader that is gaining in popularity is the memory module. This is simply an EPROM (or EEPROM) memory and associated electronic circuitry. Written and edited programs are read into the EPROM memory, which is then transported to the PC site, where its contents are unloaded into the PC. Although program loaders have been popular with small PCs, they are now gaining in popularity with larger PCs.

Some small, relatively inexpensive PCs use ROM, PROM, or EPROM chips for memory. For these PCs, a *memory burner* is used to impress a program onto the memory chips. These memories, as discussed in Chapter 6, are not as easily reprogrammed as RAM. Thus, memory burners do not find much use in PCs today.

Thumbwheel switches are convenient devices for inputting certain data to the PC. A thumbwheel switch is a 10-position switch, each position of which corresponds to a number (from 0 to 9). When a thumbwheel switch is tuned to a particular number (say, 7), it transmits the BCD code for that number (0111, in the case of the decimal number 7). Thumbwheel switches are used most often to input preset values (for timers, counters, etc.) and high and low limits. The thumbwheel switch is connected to the PC by four wires, one for each BCD bit.

DISPLAY DEVICES

In addition to serving as a programming and data entry device, the CRT is an excellent display device. Because of its large screen, many program steps and data points can be displayed at one time. Additionally, CRTs can be programmed to display process steps diagrammatically; i.e., process status can be displayed in the form of a flow diagram. The use of color terminals greatly enhances the utility of the CRT.

Two other display devices are often used. The *seven-segment display* is used to display numbers, which may correspond to speeds, times, distances, etc., depending upon the process. This is done with seven-segment LEDs (light-emitting diodes) or LCDs (liquid-crystal displays). Some PC manufacturers are providing *intelligent-alphanumeric displays*. These are programmable displays that, in addition to displaying numbers, can display messages or warnings that have been programmed into its memory.

DOCUMENTATION DEVICES

The *line printer* is the principal documentation device for PCs. Some manufacturers offer special documentation systems (for example, a system that translates ladder diagrams into written programs, and vice versa); however, the line printer is still an essential element of these systems.

Three printer characteristics are of importance: the baud rate, the buffer size, and the printing speed. The *baud rate* defines the number of bits per second of binary data that can be received (or transmitted) during serial communication between the PC and a peripheral. (Serial transmission of data will be discussed in the next section.) Baud rates vary widely. They may be as low as 50 for special-purpose, low-speed PCs, or as high as 19200 baud. The most common rates are 300, 1200, and 9600 baud. When selecting a line printer for use with a PC, care should be taken to make certain that the printer can work at the PC's particular baud rate. Usually, printer baud rates are adjustable over a small range.

A printer's *buffer* is actually a short-term memory. If the PC transmits data faster than the printer can print the data, the data not yet printed are stored temporarily in the printer's buffer. Buffer size should thus be considered along with the *printing speed* of the printer, which is given in characters printed per second (cps). Here again, printing speeds vary widely. A typical slow speed (daisy wheel) printer prints 20 characters per second, while printing speeds on the order of 200 characters per second are not unusual for dot matrix printers.

COMMUNICATIONS

There are two general schemes by which bits of data are transmitted from the peripheral to the PC, and vice versa. These are *parallel transmission* and *serial transmission*. These two schemes will be discussed below. Serial transmission is the one that is most often used in PC systems. A discussion of parallel transmission is included for completeness.

Parallel Transmission

The principles behind parallel transmission are demonstrated in Fig. 8-1. Data, in the form of binary bits, are stored in the PC logic gates. These bits are not transmitted until a transmit pulse is received by the gates. When the pulse is received, the bits are transmitted via the transmission lines to the gates in the peripheral device. Note that the transmit pulse is also sent to the peripheral. When it is received there, the gates are reset; i.e., cleared of previous data and readied to receive the transmission.

The most interesting feature of Fig. 8-1 is the number of transmission lines involved. In parallel transmission, each bit from each gate in the PC is carried by a separate line. For a large PC, this means that several lines are needed for communications between the PC and each peripheral. This is a disadvantage of parallel transmission, one which effectively restricts the use of parallel transmission to short distances (six feet or less). Another disadvantage is that since each line is dedicated, parallel interfaces are not usually interchangeable with interfaces for different peripherals.

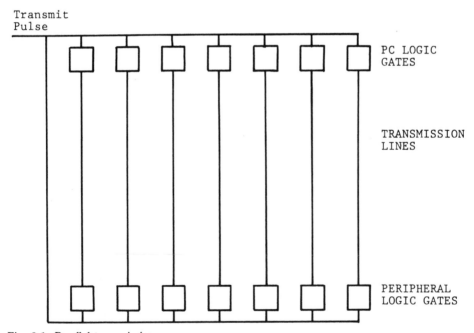

Fig. 8-1. Parallel transmission.

The primary advantage of parallel transmission is its speed. Since many bits of data are transmitted simultaneously along the several transmission lines, parallel transmission is unmatched for speed in communicating data. Unfortunately, the speed of transmission is often faster than the speed with which the peripheral can process the data. For example, a printer that prints only 20 characters per second does not require rapid transmission of data.

Note that in Fig. 8-1 a transmit pulse is used both to transmit data from the PC to the peripheral and to reset the peripheral's logic gates. Synchronization of this transmit pulse is essential. This synchronization is obtained through the use of a *clock* in the PC, which provides transmit pulses at regular intervals. Note also that the entire transmission from the PC to the peripheral occurs during one clock pulse. That is, the transmit pulse starts the transmission, and the transmission is completed during the time interval between transmit pulses.

Serial Transmission

The principles behind serial transmission are demonstrated in Fig. 8-2. Data from the PC logic gates are accepted one at a time by the driver. The acceptance of the data bits from the PC is prompted by pulses from the master timer. When each bit of data is received by the driver, it is driven, or transmitted, along a single transmission line to the receiver. The receiver directs each bit to the correct peripheral logic gate, in step with a pulse from the timer. Notice that there is a line connecting the master timer with the timer. This is to synchronize the two timers so that each received bit is directed to the correct gate.

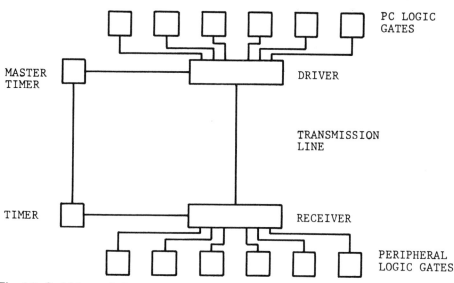

Fig. 8-2. Serial transmission.

The most interesting feature of serial transmission is the fact that only one transmission line is used. All bits of data are transmitted serially over the one line, rather than all at once over several lines as in parallel transmission. This makes serial transmission slower than parallel transmission. The advantage of serial transmission is that, since only one line is used, it allows data transmission over longer distances than are achievable with parallel transmission.

Serial transmission is used most often in PC systems because the peripherals used in PC systems are generally slow (sometimes as slow as 110 baud) and require long cable connections. A typical serial interface allows data transmission over a distance of fifty feet, compared with six feet for parallel transmission.

The driver and receiver shown in Fig. 8-2 are nothing more than combinations of AND and OR gates that are activated sequentially by timing pulses from the master timer or the timer.

In most PC systems, data are transmitted in the form of an ASCII code. Figure 8-3 shows the pulse train for the ASCII code for the comma (binary 0101100), which is compared with the pulse train from the master timer. The first pulse from the master timer signals the beginning of the transmission. The binary digits 0 and 1 are represented by zero and positive voltage levels, respectively, and are transmitted in step with the pulses from the timer. A peripheral, such as a printer, that receives the ASCII code 0101100 "recognizes" the signal as a comma. There is a seven-digit ASCII code for each number, letter, and symbol on the PC's keyboard.

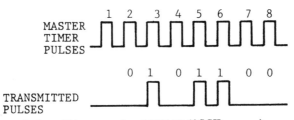

Fig. 8-3. Serial transmission of binary number 0101100 (ASCII comma).

The system shown in Fig. 8-2 allows transmission of data only from the PC to the peripheral. This is called *unidirectional*, or *simplex*, transmission. If the driver at the PC and the receiver at the peripheral are both replaced with driver/receivers (as shown in Fig. 8-4), then the form of transmission is *half-duplex*. Half-duplex transmission allows data to be transmitted from the PC to the peripheral and from the peripheral to the PC; however, data transmission occurs only in one direction at a time. In order to acquire the ability to transmit data simultaneously in two directions, the *full-duplex* system shown in Fig. 8-5 is needed. Note that the full-duplex system requires two lines for the simultaneous two-way transfer of data.

Serial Interfaces

In the discussion above, serial transmission was introduced as requiring only one transmission line. A glance at Fig. 8-5, which describes full-duplex transmission, indicates that this was a simplification. In addition to transmit and receive lines, a line is needed to synchronize the two timers. A signal ground is also needed (but not shown in any of the figures). There are several other lines that are typically contained in a serial transmission cable. These include a protective ground line and control lines, such as lines that carry signals requesting to send data, lines that carry all-clear signals for the initiation

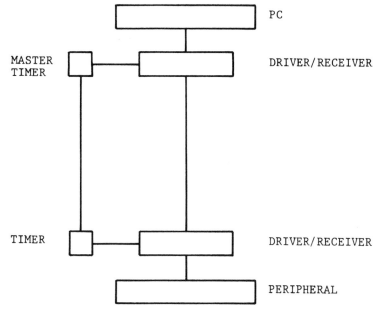

Fig. 8-4. Half-duplex transmission.

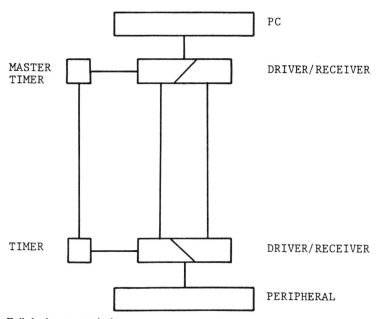

Fig. 8-5. Full-duplex transmission.

of transmission, lines that carry signals indicating that data have been received, and so on. Some standard specification of the number and physical arrangement of these transmission lines is needed in order to provide full compatibility of peripherals with various models of PCs.

There are several *communication standards* that specify signal levels and physical details of the interface, or transmission cable. The four that find most common usage in PC systems are the RS-232C, the RS-449, the 20mA current loop, and the IEEE 488. The RS-232C and RS-449 are officially proclaimed as standards by the Electronic Industries Association (EIA). The IEEE 488 is officially proclaimed as a standard by the Institute of Electrical and Electronics Engineers (IEEE). The 20mA current loop is not officially defined by any organization, but has become a standard through common usage. Of these four, the most popular in PC systems is the RS-232C.

The physical details of these standard interfaces can be obtained from the EIA or the IEEE, or from standard reference books, and will not be examined here. Note that

(a)

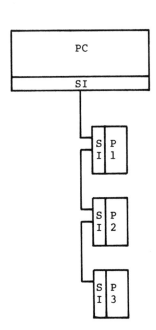

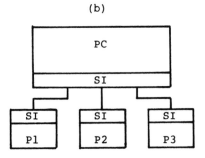

Fig. 8-6. PC-peripheral configurations: (A) Daisy chain. (B) Star. SI stands for *serial interface*. Ps represent peripherals.

even a standard such as the RS-232C interface can have many variations. For example, the full RS-232C interface contains 25 lines; however, in most instances peripherals require only three to five of the lines for proper operation.

Multiple Peripherals

Serial communication between a PC and several peripherals can be achieved by using either the *daisy chain* configuration or the *star* configuration. These are shown in Fig. 8-6.

These two configurations offer the advantages of reduced material and labor costs for large systems that have peripherals at several locations.

SUMMARY

This chapter presented the three basic classes of PC peripherals. These classes are quite similar to those associated with personal computers. The basics of parallel and serial communication between peripherals and the PC were reviewed. Of the two, serial communication is used most often in PC systems.

Chapter 9

PC Software

THE SOFTWARE THAT IS USED WITH A PROGRAMMABLE CONTROLLER IS PERHAPS THE MOST important part of the PC system for the everyday user. It is the software that allows the user to instruct the PC to perform its functions and that translates the PC's electrical signals into a form the user can understand. There are two general classes of languages used in PC software. These are *low-level* (or *basic*) languages and *high-level* languages. These will be discussed in turn.

Programming languages vary widely in their details, even though they are usually quite similar in their nature. This is especially true of the languages employed by different PC manufacturers. It is not the purpose of this chapter to familiarize you with the whole range of programming languages available today; rather, this chapter will introduce you to the fundamentals of PC programming languages. Specific examples of programming languages from certain suppliers will be used in this chapter. These will be used for illustrative purposes only.

LOW-LEVEL LANGUAGES: LADDER DIAGRAMS

In Chapter 5, relay ladder and contact ladder diagrams were introduced. The contact ladder diagram is the most widely used low-level PC language. You should now review the section of Chapter 5 entitled "LADDER DIAGRAMS."

Figure 9-1 shows the contact symbols that were introduced in Chapter 5. There are three basic symbols: the normally open (N.O.) contact symbol, the normally closed (N.C.) contact symbol, and the coil (or output) symbol. The other symbols shown in Fig. 9-1 are simply modifications of the three basic symbols.

The symbols shown in Fig. 9-1 are insufficient to program a PC in ladder diagram language. Several additional symbols are necessary. These additional symbols fall into

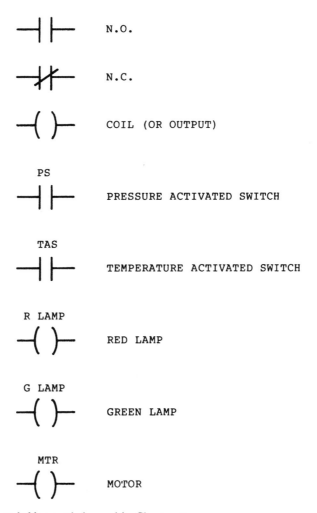

Fig. 9-1. Contact ladder symbols used in Chapter 5.

one of two categories: basic programming symbols or extended (or enhanced) programming symbols. The basic programming symbols are shown in Fig. 9-2. In addition to the normally open, normally closed, and output symbols presented previously, five other symbols are shown. The first three (out NOT, latch out, and unlatch out) are commonly known as *relay-type instructions*, because they were developed for operations with relays. The last two symbols (timer and counter) are instructions that, although basic to the operation of the PC, are not associated with relay operations.

The first three symbols in Fig. 9-2 were discussed in Chapter 5 in conjunction with relays. Since the PC is designed to replace relays in control applications, these three symbols have slightly different meanings with respect to PC programming. The symbol for the N.O. contact indicates that a specific signal is needed in order to close the contact and complete the path for the flow of current through the rung of the ladder. The contact

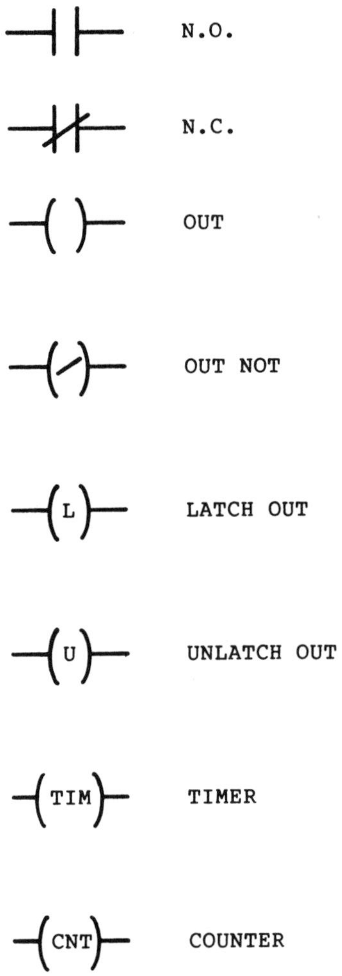

—| |—   N.O.

—|/|—   N.C.

—( )—   OUT

—(/)—   OUT NOT

—(L)—   LATCH OUT

Fig. 9-2. Basic programming symbols.

—(U)—   UNLATCH OUT

—(TIM)—   TIMER

—(CNT)—   COUNTER

symbol does not necessarily represent a contact in PC programming. It may, for example, represent an input signal. If an input signal is present, the "contact" will be closed and current will proceed to flow through the rung. In the absence of an input signal, no current will flow through the rung. The inverse holds true for the N.C. contact. The out (or output) symbol is extremely general. It may, for example, represent a lamp, motor, coil, or any other device that is energized when the current path through the rung is completed.

The three remaining relay-type instructions require some explanation, as do the timer and counter instructions.

The relationship between the out NOT and the out symbols is similar to the relationship between the normally closed and the normally open contact symbols. An out NOT symbol indicates that the output is de-energized whenever a current path through the rung is completed. In other words, the output is in the ON condition when there is no continuity through the rung and is in the OFF condition when there is continuity through the rung. This set of conditions is the inverse of what would be found for an out symbol.

The latch out and unlatch out symbols are derived from the *latching relay*. The latching relay is a retentive relay that contains two coils: the latch coil and the unlatch coil. When the latch coil is activated, the N.O. contacts close and the N.C. contacts open. These contacts remain in this condition even if the latch coil is de-energized. When power is supplied to the unlatch coil, the contacts return to their normal positions; i.e., the N.O. contacts open and the N.C. contacts close. In other words, when the latch coil of the relay is activated, the coil (for all practical purposes) behaves as if it is energized until power has been supplied to the unlatch coil. The same holds true for the unlatch coil: the relay remains in the unlatch state until power is supplied to the latch coil.

Because the latching relay contains two coils, it requires two rungs on the ladder diagram. This is shown in Fig. 9-3. In this figure, the outputs of the two rungs represent the same relay and are designated by the same number (001). When contact number 2 is closed, the output is latched. The output remains latched until contact number 3 is closed.

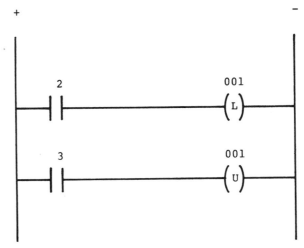

Fig. 9-3. Latch and unlatch outputs.

Latching relays are not found in PCs, of course. The latch out and unlatch out symbols merely represent PC output conditions that simulate the action of the latching relay.

The two remaining basic programming symbols are the timer and the counter. These two instructions are similar. Both are used either to energize or to de-energize a device or output after a specified time interval or count. When the timer is activated, it begins counting pulses (or time intervals) until the counted time equals some preset value. The counter, on the other hand, keeps count of events (such as pulses or signals). In the example given in Chapter 5 of the weighing platform at a shipping station, a counter could be used to keep track of the number of completely filled cartons. The preset value for this counter would be the number of cartons necessary to fill a shipping skid. Counters may either count up or count down; that is, they may count up to a preset value or count down from a preset value.

Basic programming symbols may vary from manufacturer to manufacturer. For example, some PC suppliers use a circle to denote an output, rather than the parentheses that have been used in this chapter. These differences are usually minor. The user's manual for the PC should be consulted in order to determine the appropriate programming symbols for a given PC.

The most widely used extended programming symbols are shown in Fig. 9-4. The first four symbols indicate *arithmetical operations* (addition, subtraction, multiplication, and division). These four operations can be performed on constants, on data stored in memory, or on a combination of the two. When an arithmetical operation, such as addition, is performed on two constants, the ladder diagram is similar to that shown in Fig. 9-5.

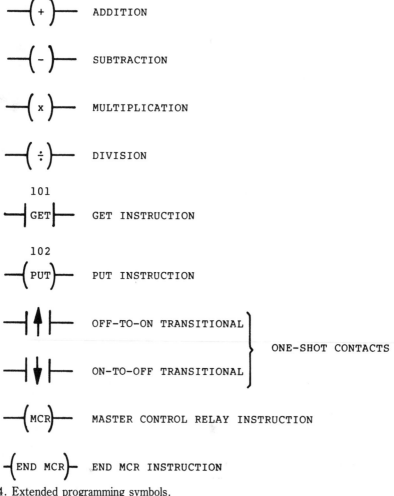

Fig. 9-4. Extended programming symbols.

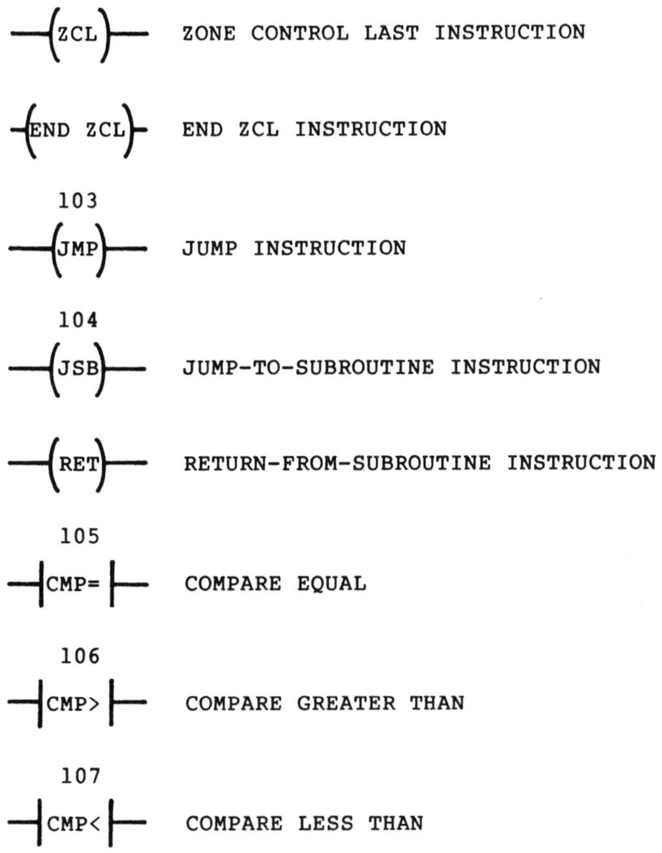
ZCL
ZONE CONTROL LAST INSTRUCTION
END ZCL
END ZCL INSTRUCTION
103
JMP
JUMP INSTRUCTION
104
JSB
JUMP-TO-SUBROUTINE INSTRUCTION
RET
RETURN-FROM-SUBROUTINE INSTRUCTION
105
CMP=
COMPARE EQUAL
106
CMP>
COMPARE GREATER THAN
107
CMP<
COMPARE LESS THAN

Fig. 9-4. (Continued.)

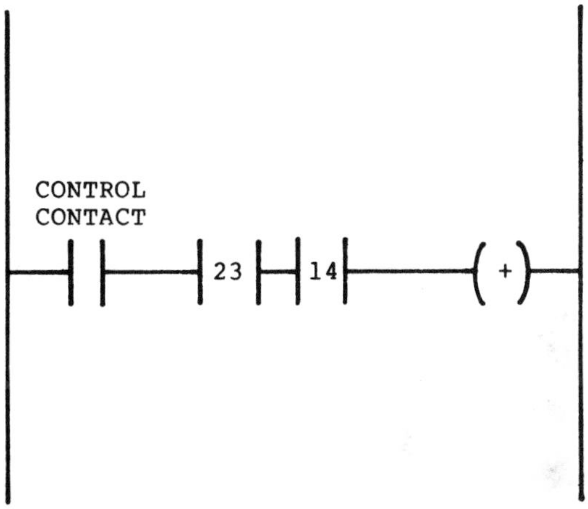
CONTROL
CONTACT
23
14
+

Fig. 9-5. Addition of two constants.

The rung in Fig. 9-5 represents the addition of the two numbers 23 and 14. The contact on the far left is called the *control contact*. When the control contact is closed, addition is performed on the two numbers.

An arithmetical operation on data stored in memory requires the fifth symbol shown in Fig. 9-4, the GET instruction. The GET instruction includes a number above the contact symbol. This number is very similar to a post office box number. It is the location or address in the PC's memory where a particular piece of data is stored. For example, a temperature sensor may be directed to store its output in memory address number 101. At a later time, the PC can access the temperature data (with the GET 101 instruction) and subtract it from a set-point value.

These memory locations for the temporary storage of data, instructions, and information are often called *registers*. A PC register normally contains space for 16 bits; that is, a 16 bit number can be stored in one register. The manufacturer's literature should be consulted to determine the register width for a particular PC.

Figure 9-6 demonstrates the subtraction of data. When the control contact is closed, the data in registers 760 and 202 are accessed, and the datum in register 202 is subtracted from the datum in register 760. The number above the subtraction symbol is the register number in which the result is stored. In this case, the difference between register 760 and register 202 is stored in register number 007.

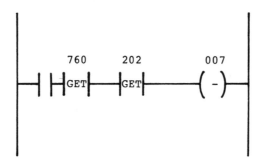

Fig. 9-6. Subtraction of data.

Operations on both data and constants can also be performed. In Fig. 9-7, the contents of register 702 are multiplied by the constant 10. The result is stored in register 521. When the control contact for rung number 2 is closed, this product is divided by the contents of register 221, and the quotient is stored in register number 336.

The companion to the GET instruction is the PUT instruction. This symbol is shown in Fig. 9-4. The PUT instruction includes a number above the coil symbol, as does the GET instruction. It is used to store outputs in a particular register or to move data from one register to another.

Two very useful instructions are the *one-shot contacts* shown in Fig. 9-4. The off-to-on transitional contact closes for one scan of the program when a triggering signal is received. The on-to-off transitional contact opens for one program scan when a triggering signal is received. Note that the contact does not remain closed (or opened,

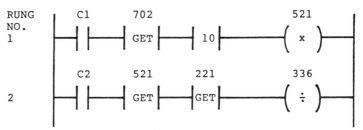

Fig. 9-7. Arithmetical operations on data and constants.

in the case of the on-to-off transitional); it simply closes for one program scan. A typical application for the one-shot would be in unlatching a latched relay when a signal is received. Any operation that should be performed only once will very likely employ the one-shot instruction.

The *master control relay (MCR)* and *END MCR* instructions shown in Fig. 9-4 are used to control the operation of the program. The MCR and END MCR statements isolate a portion of the ladder program, much in the same way that a subroutine is isolated in a computer program. This is shown in Fig. 9-8. In this figure, rungs 4 and 5 constitute the portion of the program that is isolated by the MCR (rung 3) and END MCR (rung 6) instructions.

The program works as follows. The arithmetical operations of multiplication and division are performed by rungs 1 and 2. If control contact C3 is *not* closed, rungs 3 through 6 are skipped, or ignored, and the program continues with rung 7. When the control contact for rung 7 is closed, timer number 070 begins timing in one-second intervals (BASE 01) until a preset value of 60 seconds (PRE 60) is obtained. At this point counter number 121 counts up by one unit (UP 01) from its preset value of zero (PRE 00). After the counting operation in rung 8 is performed, timer 070 is reset and once again times one-second intervals up to a preset value of 60.

If, however, control contact C3 is closed, rungs 4 and 5 are performed. Rung 6 then indicates the end of the isolated rungs, and the program proceeds to rung 7. Note that if control contact C3 is not closed, the program skips automatically from rung number 2 to rung number 7.

The group of ladder rungs controlled by the MCR and END MCR instructions is called a *zone* of ladder rungs. The MCR instruction is often used as an *override* instruction. Typically, an MCR/END MCR pair will be included in a program to protect the PC in the event of an emergency. For example, in Fig. 9-8 control contact C3 might close when current consumption by the power supply approaches a critical value. One-shot number 4 would then unlatch relay number 333, thereby protecting the system. The unlatching of relay 333 would trigger the normally closed contact in rung 5, and a critical piece of data would be obtained from register 336 and stored safely in register 402.

The MCR instruction and the ZCL instruction (which will be discussed next) are often called *skip* instructions.

The *zone control last state (ZCL)* and *END ZCL* instructions are almost identical with the MCR and END MCR instructions, respectively. The only significant difference is that if the MCR output is not activated (i.e., if the control contact for the MCR rung is not closed), the outputs within the MCR-controlled zone are de-energized. If the ZCL output is de-energized (i.e., if the control contact for the ZCL rung is not closed), the outputs within the ZCL-controlled zone are held in their last states.

The next two instructions in Fig. 9-4 are the *jump* and *jump-to-subroutine* instructions. Both of these instructions include a number above the coil symbol that indicates the contact that is the object of the instruction. When the control contact for

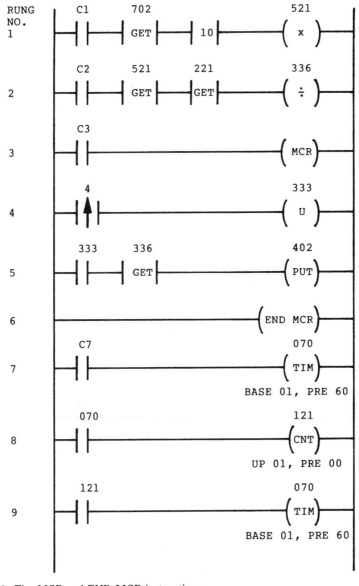

Fig. 9-8. The MCR and END MCR instructions.

a rung containing a jump instruction is closed, the microprocessor is instructed to jump to the program rung containing the control contact referenced by the jump instruction. The jump instruction is thus a method of changing the order in which a program is executed. The jump-to-subroutine instruction is equivalent to the GOSUB instruction often found in computer programs. For example, when the control contact for the ladder rung containing the jump-to-subroutine instruction is closed, the microprocessor automatically jumps to the rung containing the control contact referenced in the JSB instruction. The microprocessor then proceeds to process the subroutine.

The end of the subroutine is marked with the RET instruction (also shown in Fig. 9-4). RET stands for return-from-subroutine and directs the microprocessor to return to the program rung immediately following the jump-to-subroutine instruction. This is demonstrated in Fig. 9-9.

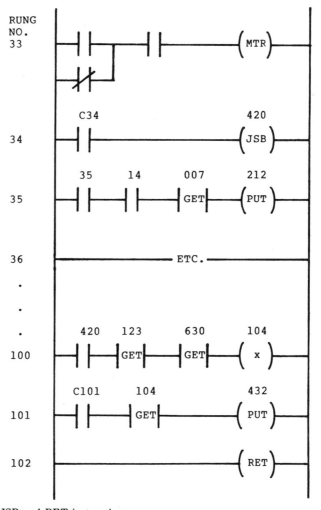

Fig. 9-9. The JSB and RET instructions.

In Fig. 9-9, the subroutine is composed of ladder rungs 100 and 101. When the JSB instruction is encountered in rung 34, program execution immediately proceeds with rung 100. When the subroutine is executed and rung 102 is encountered, the microprocessor automatically returns to rung 35 to continue with the execution of the main program.

The three remaining symbols in Fig. 9-4 are comparison operations. Note that all three symbols include numbers above them that refer to particular registers. These instructions are used in conjunction with GET instructions. They are demonstrated in Fig. 9-10. When the control contact in the first rung of the program is closed, the data in register 111 are compared with the data in register 105. If the two are equal, motor number 1 is turned on. When the control contact in rung number 2 is activated, data in register 112 are compared with data in register 106. If the value of the data in register 106 is greater than the value of the data in register 112, motor number 2 is turned on. Similarly, when the control contact in rung 3 is closed, data in register 113 are compared with data in register 107. If the value of the data in register 107 is less than the value of the data in register 113, motor number 3 is turned on.

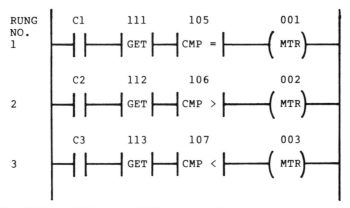

Fig. 9-10. The CMP =, CMP >, and CMP < instructions.

There are many more ladder symbols that are used in the programming of PCs. The ones presented above are the symbols most often encountered in PC programming. Of course, not every manufacturer uses the same symbols. This is shown in Fig. 9-11, which is the list of programming instructions for the SYSMAC-C120, C250, and C500 series of programmable controllers produced by Omron. Some differences between the symbols presented in Fig. 9-4 and the symbols presented in Fig. 9-11 are apparent. For example, the SYSMAC language uses a circle (and not parentheses) to indicate an output. Also, the SYSMAC symbols for the counter, timer, jump, latching relay, arithmetical operations, and comparison are different from those presented in Fig. 9-4. Although these symbols are different, consultation with the supplier's user's manual is usually sufficient to highlight the meaning of each of the individual programming symbols.

■ LIST OF INSTRUCTIONS
● Basic Instruction

Instruction	Symbol	Instruction word length (byte)	Mnemonic / Operand	Function	Data	Flag*
LOAD	⊣├─	4	LD / Relay No.	Logical start operation	Relay No.	
LOAD NOT	⊣╫─	4	LD NOT / Relay No.	Logical NOT start operation	Input/output relays Internal auxiliary relays 0000 to 6307	
AND	─┤├─	3	AND / Relay No.	Logical AND operation	Link relays LR0000 to 3115	
AND NOT	─╫─	3	AND NOT / Relay No.	Logical AND NOT operation	Holding relays HR0000 to 3115	
OR	┤├	3	OR / Relay No.	Logical OR operation	Timers TIM000 to 127 Counters CNT000 to 127	
OR NOT	╫	3	OR NOT / Relay No.	Logical OR NOT operation	Temporary memory relays TR0 to 7 (Temporary memory relays can only be used with the LD instruction.)	No change
AND LOAD		5	AND LD	Logical AND operation with the previous condition	—	
OR LOAD		5	OR LD	Logical OR operation with the previous condition		
OUT	─◯	6	OUT / Relay No.	Outputs the result of a logical operation to the specified output relay, internal auxiliary relay, latching relay, or shift register.	Relay No. 000 to 6015 LR0000 to 3115 HR0000 to 3115 TR0 to 7 (Temporary memory relays can be used with the OUT instruction only.)	
OUT NOT	─⊘	8	OUT NOT / Relay No.	Inverts the result of a logical operation and then outputs it to a specified output relay, internal auxiliary relay, holding relay, or shift register.		
TIMER	─(TIM)	8	TIM / Timer No. / Set value	ON-delay timer operation Set time: 0 to 999.9s	Timers/counters 000 to 127 Set value	(ER)
COUNTER	CP R CNT	8	CNT / Counter No. / Set value	Down counter operation Set value: 0 to 9999	Constant 0000 to 9999 External setting 00 to 63 LR00 to 31 HR00 to 31	

NOTE: * (<) : 6307
(=) : 6306
(>) : 6305
(CY) : 6304
(ER) : 6303

INFORMATION COURTESY OF OMRON ELECTRONICS, INC.

Fig. 9-11. Programming instructions for the SYSMAC C120, C250, and C500 series PCs.

PARTS SHARED BY SYSMAC-C500, SYSMAC-C250, AND SYSMAC-C120

● Applied Instructions

FUN No.	Instruction	Symbol	Instruction word length (byte)	Mnemonic Operand	Function	Data	Flag
00	NO FUNC- TION		3	NOP (FUN00)	−	−	
01	END	END	3	END (FUN01)	End of a program		
02	INTER- LOCK	IL	4	IL (FUN02)	Causes all the relay coils between this instruc- tion and the ILC instruction to be reset or not reset according to the result immediate- ly before this instruc- tion.	−	
03	INTER- LOCK CLEAR	ILC	4	ILC (FUN03)	Clears the IL instruction.		
04	JUMP	JMP	4	JMP (FUN04)	Causes all the contents of a program between this instruction and the JME instruction to be ignored or executed according to the result immediately before this instruction.	−	
05	JMP END	JME	4	JME (FUN05)	Clears the JMP instruction.		No change
06	FAIL RESET	FAL 00	6	FAL (FUN06) 00	Clears FAL or FALS instructions or alarm indications.	Content of an alarm is cleared at one scanning.	
06	DIAG- NOSTIC (FAL)	FAL	6	FAL (FUN06) No.	Indicates the type of failure or abnormal mode.	No. 01 to 99	
07	DIAG- NOSTIC (FALS)	FALS	6	FALS (FUN07) No.	Indicates the type of failure or abnormal mode that causes the PC to stop operation.		
10	SHIFT REGIS- TER	IN CP R SFT	8	SFT (FUN10) Start CH-No. CH-No. End	Shift register operation 15 0 15 0 End CH Start CH --IN	Channel Numbers 00 to 60 LR00 to 31 HR00 to 31 * Start CH ≤ End CH * Start and end channels can be used at same area.	
11	LATCH- ING RELAY	S KEEP R	6	KEEP (FUN11) Relay No.	Latching relay operation	Relay No. 0000 to 6015 LR0000 to 3115 HR0000 to 3115	
12	UP-DOWN COUN- TER	ACP SCP R CNTR	8	CNTR (FUN12) Counter No. Set value	Up-down counter operation Set value: 0000 to 9999	Timer/counters 000 to 127 Set value Constant 0000 to 9999 External setting 00 to 63 LR00 to 31 HR00 to 31	(ER)

Fig. 9-11. (Continued.)

FUN No.	Instruction	Symbol	Instruction word length (byte)	Mnemonic	Operand	Function	Data	Flag
13	DIFFER-ENTIA-TION UP	DIFU	6	DIFU (FUN13)	Relay No.	Causes a specified relay to operate for one scan time at the leading edge of the result of a logical arithmetic operation.	Relay No. 0000 to 6015 LR0000 to 3115 HR0000 to 3115	No change
14	FIFFER-ENTIA-TION DOWN	DIFD	6	DIFD (FUN14)	Relay No.	Causes a specified relay to operate for one scan time at the trailing edge of the result of a logical arithmetic operation.		
15	HIGH-SPEED TIMER	TIMH	8	TIMH (FUN15)	Timer No. Set value	Performs a high-speed on-delay (down type) timer operation. Set time: 00.00 to 99.99 sec	Timers/counters 000 to 127 Set value Constant 0000 to 9999 External setting 00 to 63 LR00 to 31 HR00 to 31	(ER)
16	WORD SHIFT	WSFT	8	WSFT (FUN16)	D1 D2	Shifts words by I/O channel data unit (i.e., 16 bits). Data "0" → D1 ... D2	D1 D2 Same as MOVE instruction * Start CH ≤ End CH * Start and end channels can be used at same area.	
20	COM-PARE	CMP	8	CMP (FUN20)	S1 S2	Compares a channel data or a 4-digit constant against another channel data. $S_1 \leqq^< _> S_2$	S . S1 . S2 Input/output relays Internal auxiliary relays 00 to 63 Link relays LR00 to 31 Holding relays HR00 to 31 Timers TIM000 to 137 Counters CNT000 to 127 Data memory DM000 to 511 Indirect addressed *DM000 to 511 Constant 0000 to FFFF	(<)(=) (>) (ER)
21	MOVE	MOV	8	MOV (FUN21)	S D	Transfers a channel data or a 4-digit constant (16 bits) to a specified channel. S → D		
22	MOVE NOT	MVN	8	MVN (FUN22)	S D	Inverts a channel data or a 4-digit constant and transfers it to a specified channel. $\bar{S}$ → D	D 00 to 60 LR00 to 31 HR00 to 31 DM000 to 511 *DM000 to 511	
23	BCD-TO-BIN CON-VERSION (BIN)	BIN	8	BIN (FUN23)	S D	Converts BCD data into binary data. S → D (BCD) (BIN)	S 00 to 63 LR00 to 31 HR00 to 31 TIM/CNT000 to 127 DM000 to 511 *DM000 to 511 * TIM/CNT can be used with BCD-TO-BIN conversion instruction only. D Same as MOV instruction.	(=) (ER)
24	BIN-TO-BCD CON-VERSION (BCD)	BCD	8	BCD (FUN24)	S D	Converts binary data into BCD data. S → D (BIN) (BCD)		

Fig. 9-11. (Continued.)

PARTS SHARED BY SYSMAC-C500, SYSMAC-C250, AND SYSMAC-C120

FUN No.	Instruction	Symbol	Instruction word length (byte)	Mnemonic Operand	Function	Data	Flag
25	ARITH-METIC SHIFT LEFT	ASL	7	ASL (FUN25) D	Shifts a channel data including a carry to the left.	D 00 to 60 LR00 to 31 HR00 to 31 DM000 to 511 *DM000 to 511	
26	ARITH-METIC SHIFT RIGHT	ASR	7	ASR (FUN26) D	Shifts a channel data including a carry to the right.		(=) (CY) (ER)
27	ROTATE LEFT	ROL	7	ROL (FUN27) D	Rotates a channel data left, including a carry.		
28	ROTATE RIGHT	ROR	7	ROR (FUN28) D	Rotates a channel data right, including a carry.		
29	COMPLE-MENT	COM	7	COM (FUN29) D	Inverts a channel data (16-bit) $\bar{D} \rightarrow D$		(=) (ER)
30	ADD	ADD	10	ADD (FUN30) S1 S2 D	Performs BCD addition of a channel data or a 4-digit constant to a specified channel data. $S_1 + S_2 + \boxed{CY}$ $= D, \boxed{CY}$	S_1 S_2 00 to 63 LR00 to 31 HR00 to 31 TIM/CNT000 to 127 DM000 to 511 *DM000 to 511 Constant 0000 to 9999 D Same as MOVE instruction	(=) (CY) (ER)
31	SUB-TRACT	SUB	10	SUB (FUN31) S1 S2 D	Performs BCD subtraction of a channel data or a 4-digit constant from a specified channel data. $S_1 - S_2 - \boxed{CY}$ $= D, \boxed{CY}$		
32	MULTI-PLY	MUL	10	MUL (FUN32) S1 S2 D	Performs BCD multiplication of a channel data by a channel data or a 4-digit constant. $S_1 \times S_2 = $ D , D+1 (LSB) (MSB)	S_1 S_2 Same as ADD instruction. D 00 to 59 LR00 to 30 HR00 to 30 DM000 to 510 *DM000 to 511	
33	DIVIDE	DIV	10	DIV (FUN33) S1 S2 D	Performs BCD division of a channel data by a specified channel data or a 4-digit constant. $S_1 \div S_2$ $= D$ Remainder D+1		(=) (ER)
34	AND WORD	ANDW	10	ANDW (FUN34) S1 S2 D	Performs a logical AND operation between two 16-bit data. $S_1 \wedge S_2 \rightarrow D$	S_1 S_2 Same as MOVE instruction. D Same as MOVE instruction.	
35	OR WORD	ORW	10	ORW (FUN35) S1 S2 D	Performs a logical OR operation between two 16-bit data. $S_1 \vee S_2 \rightarrow D$		

Fig. 9-11. (Continued.)

PARTS SHARED BY SYSMAC-C500, SYSMAC-C250, AND SYSMAC-C120

FUN No.	Instruction	Symbol	Instruction word length (byte)	Mnemonic / Operand	Function	Data	Flag
36	EX-CLUSIVE OR WORD	XORW	10	XORW (FUN36) / S1 / S2 / D	Performs an exclusive logical OR operation between two 16-bit data. $S_1 \veebar S_2 \rightarrow D$		
37	EX-CLUSIVE OR NOT WORD	XNRW	10	XNRW (FUN37) / S1 / S2 / D	Performs an exclusive logical OR NOT operation between two 16-bit data. $S_1 \veebar S_2 \rightarrow D$	—	(=) (ER)
38	INCRE-MENT (INC)	INC	7	INC (FUN38) / D	Increments a 4-digit BCD data by 1. $D + 1 \rightarrow D$		
39	DECRE-MENT (DEC)	DEC	7	DEC (FUN39) / D	Decrements a 4-digit BCD data by 1. $D - 1 \rightarrow D$		
40	SET CARRY (STC)	STC	4	STC (FUN40)	Sets the carry (CY) to "1". $1 \rightarrow$ CY		No change
41	CLEAR CARRY (CLC)	CLC	4	CLC (FUN41)	Clears the carry (CY) to "0". $0 \rightarrow$ CY	—	

● **Special Instructions**

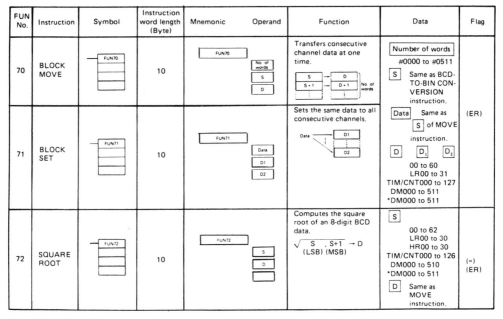

FUN No.	Instruction	Symbol	Instruction word length (Byte)	Mnemonic / Operand	Function	Data	Flag
70	BLOCK MOVE	FUN70	10	FUN70 / No. of words / S / D	Transfers consecutive channel data at one time.	Number of words #0000 to #0511 — S Same as BCD-TO-BIN CONVERSION instruction.	(ER)
71	BLOCK SET	FUN71	10	FUN71 / Data / D1 / D2	Sets the same data to all consecutive channels.	Data Same as S of MOVE instruction. D D₁ D₂ 00 to 60, LR00 to 31, TIM/CNT000 to 127, DM000 to 511, *DM000 to 511	
72	SQUARE ROOT	FUN72	10	FUN72 / S / D	Computes the square root of an 8-digit BCD data. $\sqrt{\ S\ ,\ S+1\ } \rightarrow D$ (LSB) (MSB)	S 00 to 62, LR00 to 30, HR00 to 30, TIM/CNT000 to 126, DM000 to 510, *DM000 to 511 D Same as MOVE instruction.	(=) (ER)

Fig. 9-11. (Continued.)

PARTS SHARED BY SYSMAC-C500, SYSMAC-C250, AND SYSMAC-C120

FUN No.	Instruction	Symbol	Instruction word length (byte)	Mnemonic	Operand	Function	Data	Flag
73	DATA EX-CHANGE	FUN73	10	FUN73	D1 D2 —	Exchanges data between channels. D1 ↔ D2	D_1 D_2 Same as BLOCK MOVE instruction.	
74	ONE DIGIT SHIFT LEFT	FUN74	10	FUN74	D1 D2 —	Shifts data between the start and end channels to the left by 1 digit (i.e., 4 bits). D_1 … 0 D_{1+1} D_2	S Same as BCD-TO-BIN CONVERSION instruction. D , D_1 , D_2 Same as MOVE instruction. Digit designation #0000 to #0003	
75	ONE DIGIT SHIFT RIGHT	FUN75	10	FUN75	D1 D2 —	Shifts data between the start and end channels to the right by 1 digit (i.e., 4 bits). D_1 D_{1+1} 0 → D_2		
76	4-TO-16 DE-CODER	FUN76	10	FUN76	S Digit designation D	Decodes a 4-bit binary data of 16 bits to a 16-bit channel data. S #3 #2 #1 #0 0 to F D Only 1 bit is 1		
77	16-TO-4 EN-CODER	FUN77	10	FUN77	S D Digit designation	Encodes a 16-bit decimal data into 4 bits of another 16-bit binary data. S MSB 15 … LSB 0 1 of MSB side D #3 #2 #1 #0 0 to F		(ER)
78	7-SEG-MENT DE-CODER	FUN78	10	FUN78	S Digit designation D	Converts 4 bits of 16-bit data into an 8-bit data for 7-segment display. S #3 #2 #1 #0 0 to F D 8 bit data		
79	FLOAT-ING POINT DIVIDE	FUN79	10	FUN79	S1 S2 D	Performs a floating-point arithmetic operation between two 7-digit BCD data. (LSB) (MSB) (S1, S1 + 1) ÷ (S2, S2 + 1) → (D, D + 1)	S1 S2 Same as S of SQUARE instruction D Same as MULTIPLY instruction	
80	DATA DISTRI-BUTION	FUN80	10	FUN80	S D1 D2	Distributes 16-bit data to a channel that is a base address with an offset added. S → Base address (D1) + Offset (D2) (S) → (D1 + D2)	S Same as S of MOVE instruction D1 D2 Same as D of BLOCK MOVE instruction	

Fig. 9-11. (Continued.)

PARTS SHARED BY SYSMAC-C500, SYSMAC-C250, AND SYSMAC-C120

FUN No.	Instruction	Symbol	Instruction word length (byte)	Mnemonic	Operand	Function	Data	Flag
81	DATA EXTRAC-TION	FUN81	10	FUN81	S1, S2, D	Extracts 16-bit data from a channel that is a base address with an offset added and transfers the data to a specific channel. 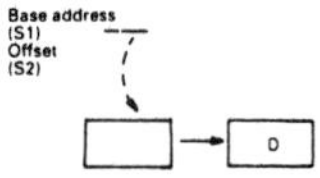	S1: 00 to 63, LR00 to 31, HR00 to 31, TIM/CNT000 to 127, DM000 to 511, *DM000 to 511; S2 D: Same as D of BLOCK MOVE instruction	
82	BIT TRANS-FER	FUN82	10	FUN82	S, Control data, D	Transfers a specific bit to another specified bit. 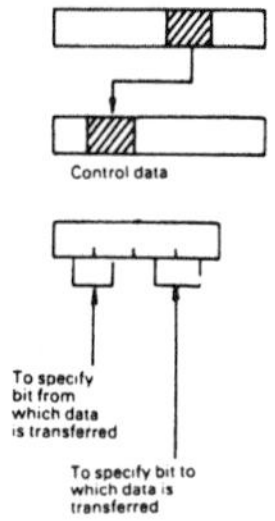	S: 00 to 63, LR00 to 31, HR00 to 31, DM000 to 511, *DM000 to 511, Constant 000 to FFFF; Control data: 00 to 60, LR00 to 31, HR00 to 31, TIM/CNT000 to 127, DM000 to 511, *DM000 to 511, Constant; D: Same as D of MOVE instruction	(ER)
83	DIGIT TRANS-FER	FUN83	10	FUN83	S, Control data, D	Transfers channel data in units of digits (4 bits) to a specified channel. 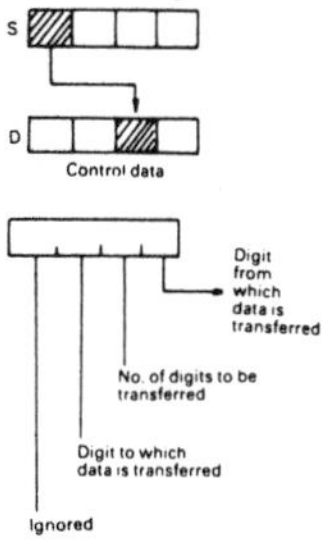	S: Same as S of ADD instruction; Control data: 00 to 60, LR00 to 31, HR00 to 31, TIM/CNT000 to 127, DM000 to 511, *DM000 to 511, Constant; D: Same as D of BLOCK SET instruction	
84	LEFT/RIGHT SHIFT REGIS-TER	FUN84	10	FUN84	Control data, Start CH No, End CH No	Shifts specified 16-bit data 1 bit to the left or to the left or to the right. 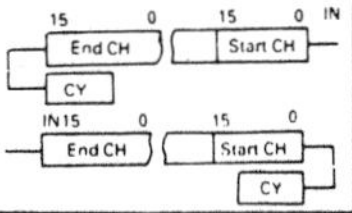	00 to 60, LR00 to 31, HR00 to 31	

Fig. 9-11. (Continued.)

PARTS SHARED BY SYSMAC-C500, SYSMAC-C250, AND SYSMAC-C120

FUN No.	Instruction	Symbol	Instruction word length (byte)	Mnemonic	Operand	Function	Data	Flag
85	TABLE COM-PARE	FUN85	10	FUN85	S T D	Compares 16-bit data against data in 16 channels (table) and outputs the results to a specified channel.	Control data CH No. 00 to 60 LR00 to 31 HR00 to 31 DM000 to 511 *DM000 to 511 • Start CH $\leq$ End CH • The start CH must be in the same relay area as the end CH.	
94	WATCH-DOG TIMER SETTING	FUN94	10	FUN94	No. of times — —	Refreshes the set time of the watchdog timer.	No. of times 0 to 63	(ER)
99	RUN STOP	FUN99	10	FUN99	— — —	Stops the RUN operation when the specified relay is turned OFF and starts the operation when it is turned ON.		

Fig. 9-11. (Continued.)

LOW-LEVEL LANGUAGES: BOOLEAN LANGUAGE

Boolean algebra was discussed in Chapter 3. Boolean language is based upon the same principles as Boolean algebra. Boolean language is a *mnemonic* language. This means that the Boolean operators (AND, OR, NOT, NAND, and NOR) are used in symbolic form to provide a programming language entirely equivalent to the ladder language.

Boolean mnemonic language is the second most popular programming language, behind ladder diagram programming. The reason for its popularity is that it is interchangeable with ladder programming. This can be seen in Fig. 9-12. The familiar ladder diagram symbols in Fig. 9-12 are equivalent to the Boolean symbols, which are composed of two to six characters. In many cases, the Boolean characters are simply the same as the ladder symbols, minus the parentheses and contact symbols. This is especially true of the timer, counter, compare, jump, jump-to-subroutine, and master control relay instructions.

Five of the symbols in Fig. 9-12 deserve special mention. These are the OR, NOR, OUT NOT, LOAD, and LOAD NOT instructions. The OR instruction is simply a N.O.

contact in parallel with a rung. Continuity through the rung can be achieved by the closing of the contacts in the rung, OR by the closing of the parallel contact. The NOR (NOT OR) instruction is similar, except that it employs a N.C. contact. The OUT NOT is similar to the OUT symbol, with one exception. An OUT symbol is energized when continuity through the rung is achieved. An OUT NOT symbol is *de-energized* when continuity through the rung is achieved. The LOAD and LOAD NOT instructions are used to symbolize the initiation of the rung, with N.O. or N.C. control contacts, respectively.

Returning to the SYSMAC programming symbols shown in Fig. 9-11, you can see that the Boolean mnemonics are given in the columns entitled "Instruction," and the ladder symbols are given in the columns entitled "Symbol." The columns entitled "Mnemonic" and "Operand" show the SYSMAC terminal keys to be pushed in order to implement the Boolean or ladder program step. (Not all of the symbols in Fig. 9-11 are ladder symbols. Some are block diagram symbols, which will be discussed in the

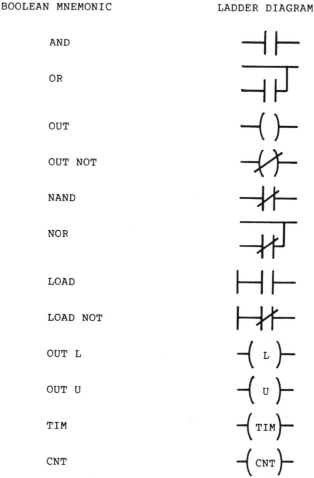

Fig. 9-12. Comparison of Boolean mnemonic and ladder diagram languages.

BOOLEAN MNEMONIC	LADDER DIAGRAM
ADD	—(+)—
SUB	—(-)—
MUL	—(x)—
DIV	—(÷)—
CMP =	—(CMP =)—
CMP >	—(CMP >)—
CMP <	—(CMP <)—
JMP	—(JMP)—
JSB	—(JSB)—
MCR	—(MCR)—
END	—(END MCR)—

Fig. 9-12. (Continued.)

next section.) Note that the SYSMAC instructions contain many Boolean and ladder instructions that have not been covered in this chapter. The meaning and usage of these instructions can be discerned from a study of the PC's user's manual.

HIGH-LEVEL LANGUAGES: BLOCK DIAGRAMS

A low-level language, such as the ladder diagram or Boolean mnemonic symbol languages discussed above, has limited applicability when you are programming complex routines. In order to implement fairly complex programs (programs which, for example, include proportional-integral-derivative control), a *high-level language* is needed. Block diagrams can be used to form a high-level language that allows the programmer to implement complex routines using the ladder diagram format.

The block diagram symbols most often encountered are shown in Fig. 9-13. These symbols are used in conjunction with the ladder symbols for the N.O. and N.C. contacts, and the OUT and OUT NOT output conditions. Block diagram language uses the ladder format; that is, the block symbols are placed in rungs and require energizing signals.

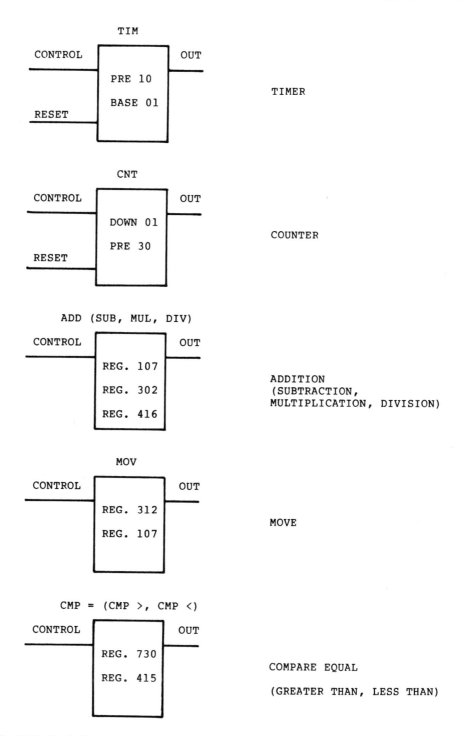

Fig. 9-13. Block diagram language symbols.

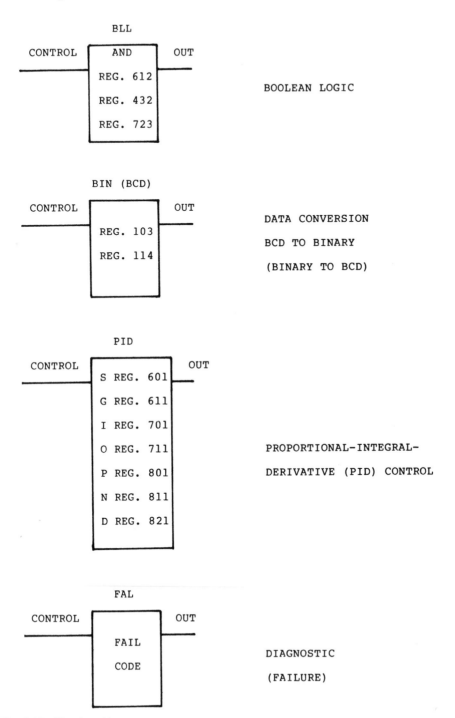

Fig. 9-13. (Continued.)

The first two symbols in Fig. 9-13, the timer and counter block symbols, illustrate the general form that block diagrams assume. Two inputs are shown. The control input is in series with a control contact. When the control contact is closed, the timer begins timing and the counter begins counting. Within the block, the preset and base values are shown. The counter includes a direction: in this case, the counter counts down from a preset value of 30 in increments of one each time the control contact is closed. The base and preset values have the same meaning as in Fig. 9-8. For example, when the control contact in the timer rung is closed, the timer begins timing in 1 second intervals up to the preset value of 10 seconds. At 10 seconds, the output goes ON or TRUE (i.e., the output is energized), and continuity through the rung is assured.

The reset inputs are not always shown in block symbols. For example, the timer symbol in Fig. 9-11 does not include a reset input. The counter symbol in Fig. 9-11, however, does include a reset input. If the reset input is present, it is used to reset the timer or counter to the OFF or FALSE (i.e., de-energized) state. Thus, a reset signal would reset the timer in Fig. 9-13 to zero and would reset the counter in Fig. 9-13 to 30. When the reset input is not present in the timer symbol, it is assumed that the timer automatically resets when its control contact is closed to begin a subsequent timing cycle.

It should be emphasized that the counter in Fig. 9-13 counts down from the preset value in increments of one. In other words, when the control contact is closed, the count value drops to 29. The count value remains at 29 *until the control contact is again closed*. In order to count down from 30 to zero, the control contact must be closed 30 times. Some counter symbols include a second output line that indicates the running count value.

The third symbol in Fig. 9-13 is the block language symbol for arithmetical operations (addition, subtraction, multiplication, and division). When the control contact is closed (so that the control input is energized), the number contained in register 107 (which can be either data or a constant) and the number contained in register 302 are used to perform an arithmetical operation. If the block symbolizes addition, the two numbers are added. If the block symbolizes subtraction, the number in register 302 is subtracted from the number in register 107. The same holds true for multiplication and division. (In the case of division, the number in register 107 is divided by the number in register 302.) The result of the operation, that is, the sum, difference, product, or quotient, is then stored in register number 416.

The output line to the right of the arithmetical operation block is not really needed, since the result of the operation is stored in a register location. The output line is generally used as an *overflow indicator*. Whenever the result of the operation is greater than the register can hold, the output line is energized. If a warning light is connected to the output, the light would signal the programmer that the capacity of the register has been exceeded. A typical overflow condition occurs when two eight-bit numbers are added to produce a nine-bit sum with a register that is only eight bits wide. This type of overflow can be corrected by a process known as *scaling*. The PC's user's manual should be consulted for details on the scaling process.

The block symbols for addition, subtraction, multiplication, and division in Fig. 9-11 (for the SYSMAC series of PCs) are very similar to those shown in Fig. 9-13. Note that the SYSMAC block language includes a square root instruction. In general, any arithmetical operation can be represented in the block form described above.

The *move* instruction in Fig. 9-13 is the equivalent of the ladder instructions GET and PUT. When the control contact is closed, the number in register 312 is acquired (GET) and moved (PUT) to register 107. When this operation is completed, the output is energized.

The move symbol in Fig. 9-13 is quite similar to the SYSMAC move symbol (Fig. 9-11). Note, however, that the SYSMAC instruction set contains other data transfer instructions in addition to the move instruction. These include shift register, word shift, move not, arithmetic shift left and shift right, rotate left and right, block move, block set, bit transfer, digit transfer, and left/right shift register. The details of these data transfer operations can be obtained from the PC's user's manual. Of the operations given above, those most commonly found in PCs are the shift register and block move. The shift register instruction simply shifts words within a *stack* of registers each time an operation is performed. (Register stacks are found in the data section of application memory. See Chapter 6.) The block move instruction is a convenient method of transferring data from one group or stack of registers to another group or stack of registers at one time. Its use avoids the repetitive use of the move instruction.

The three compare instructions (CMP =, CMP >, CMP <) can be represented in block form (Fig. 9-13). Numbers in the two registers are compared, and if the equality (or inequality) condition is satisfied, the output is energized. The SYSMAC compare instruction (Fig. 9-11) is quite similar to the one shown in Fig. 9-13.

The Boolean logic (BLL) block symbol performs the Boolean operations discussed in Chapters 2 and 3: AND, OR, NOT, NAND, NOR, and XOR. The operation to be performed is specified within the block. Thus when the control input is ON, the symbol shown in Fig. 9-13 performs the AND operation on register numbers 612 and 432, and stores the result in register number 723. The output is energized when the operation is performed.

Not all PCs have sufficient memory capacity to allow the storage of the results of block operations in separate registers. If this is the case, the result of a block operation is stored in one of the input registers, thereby erasing the input value. For example, if the BLL block in Fig. 9-13 contains only register numbers 612 and 432, the result of the AND operation will be stored in register 432. The data previously stored there would be erased.

The three remaining blocks shown in Fig. 9-13 have no analogs in either the ladder or Boolean mnemonic languages, and demonstrate the advantage of a high-level language. These symbols are data conversion, proportional-integral-derivative (PID) control, and diagnostic.

Although the data conversion block in Fig. 9-13 shows the conversion of binary-coded decimal (BCD) into binary (or binary to BCD), other conversions can occur. For example, the SYSMAC instruction set includes the conversion of 16-bit data into 8-bit data for 7-segment displays and the complement, which inverts data, in addition to the BCD-to-binary and binary-to-BCD instructions. The operation of the block in Fig. 9-13 is simple. When the control input is ON, BCD data in register 103 are converted to binary and stored in register 114. When this operation is completed, the output turns ON.

Proportional-integral-derivative (PID) control was discussed in Chapter 7 (under "Special I/O Interfaces"). That chapter emphasized the use of intelligent I/O interfaces to perform PID control. PID control can also be programmed into the PC's memory using the symbol shown in Fig. 9-13.

The set point for the process (S) and desired-gain (G) values are entered into registers 601 and 611, respectively. Input from the process (I) and the output control signal (O) use registers 701 and 711, respectively. The three remaining registers (801, 811, 821) hold the proportional (P), integral (N), and derivative (D) terms, respectively. These terms make up the PID control signal. When the control line is ON, the output line carries the PID control signal.

The absence of a PID programming symbol from an instruction set does not necessarily mean that the PC cannot perform PID control. PID is a popular control scheme, and many PC manufacturers include the PID scheme in a module that is connected to the PC. In such a case, PID control is initiated with a simple ladder output instruction in the main program.

The diagnostic block shown in Fig. 9-13 is similar to the SYSMAC diagnostic block (Fig. 9-11). When energized, this particular diagnostic symbol indicates, via a failure code, the type or mode of failure that has occurred. If, for example, an interruption in power supply to a portion of the system has occurred, the diagnostic program is energized. The program then determines that the type of failure was a power interrupt, and sends an output signal (failure code) to an output display (such as two 7-segment displays) in the diagnostic block's output line. The two-digit number displayed by the 7-segment displays alerts the operator that a failure has occurred. The operator compares the two-digit number with a list of failure codes and determines that a power interrupt was the type of failure suffered.

There are a wide variety of diagnostics available with various PCs. The PC's user's manual should be consulted for further details.

HIGH-LEVEL LANGUAGES: COMPUTER-TYPE LANGUAGES (CTL)

Users of personal computers will be familiar with the types of high-level languages discussed in this section. *Computer-type languages (CTL)* are languages that employ English statements and instructions. They are usually similar to BASIC, the widely used personal computer programming language. In fact, Encoder Products Co. manufactures two PCs (models 7152 and 7252) which use BASIC as the programming language.

Although similar to BASIC, most CTLs are easier to use than BASIC; i.e., they are more *user-friendly* or *operator-oriented*. Since CTLs offer essentially the complete programming flexibility of a computer language, their use enhances the computational and control power of the PC. And since CTLs are based on the English language, they are easier for the operators to understand.

The specifics of a particular CTL can be obtained from the PC's user's manual. Since most CTLs are quite similar to BASIC, we will examine the BASIC instruction set.

The BASIC instruction set is shown in Table 9-1. Although these instructions do not correspond exactly to ladder symbols, they do display similarities. The first four instructions, LET, INPUT, READ, and DATA, are similar in nature to the ladder contact symbols. The LET instruction is used to assign a number value to a variable. Thus if a contact is represented by the variable x, the statement "LET X = 1" would indicate that the contact is closed. The instruction INPUT accepts data from the PC's keyboard. A program containing the statement "INPUT X" would cause a prompt (such as a question mark) to be displayed on the terminal screen (or CRT). The program will not continue until the value of x is entered. The READ instruction is similar to the INPUT instruction, except that it directs the microprocessor to accept data from inputs to the PC or from DATA statements. For example, "READ T" would direct the microprocessor to accept a temperature input from a thermocouple. A DATA statement is a line in a program that contains data to be used in the execution of the program (e.g., "DATA 73.46, 892.10, 107.62"). Data in the DATA statement are accessed by the READ instruction.

The IF and THEN instructions are used together and simulate the output ladder symbol. Suppose that we wish to simulate a ladder rung that contains one N.O. contact (x) and one output (y). When the N.O. contact is closed, the output is energized. This can be simulated using the following program:

```
10   READ X
20   IF X=0 THEN Y=0
30   IF X=1 THEN Y=1
```

Table 9-1. BASIC Instruction Set.

LET	IF	TIMER ON	WAIT
INPUT	THEN	TIMER OFF	FOR
READ		TIMER STOP	STEP
DATA			NEXT
+	PEEK	GOTO	=
−	POKE	GOSUB	>
*		RETURN	<
/			
SQR			
LOG			
SIN			
COS			
etc.			
ERL	PRINT	REM	END
ERR			RUN

In other words, the IF/THEN pair establishes a condition that, when satisfied, results in a specific output. Note that each line in the program given above is numbered for programming convenience.

Timer operation is controlled by the TIMER ON, TIMER OFF, and TIMER STOP instructions. Note that timer instructions vary widely among different versions of BA-SIC. In this version, these instructions turn on, turn off, and temporarily halt the timer, respectively.

The WAIT instruction and the FOR, STEP, and NEXT instructions can be used to simulate the operation of a counter. The WAIT instruction halts the execution of a program until specified inputs are provided. Suppose that we have written a program that controls the sealing of a carton. We do not want the program to begin functioning until ten objects are placed in the carton. By using a light beam and photoelectric cell, we can obtain an electrical signal every time an object reaches the carton. If this input signal is connected to I/O terminal number 16 on the PC, we would use the following line in the control program:

```
80 WAIT 16, 10
```

This line (line number 80 in the program) simply delays the execution of the remainder of the program until 10 signals have been received at I/O terminal number 16.

The FOR, STEP, and NEXT instructions are less efficient in simulating counters, but they are very useful in other applications. FOR and STEP are used as a pair in the same program line. These three instructions allow an operation to be performed a preset number of times, and allow the setting of initial and final conditions. A typical usage is as follows:

```
10   FOR X=0 TO 10 STEP 1
20   Y=2*X
30   PRINT "X=",X,"Y=",Y
40   NEXT X
```

In this program x is varied from zero to ten in increments ("steps") of one (i.e., 0, 1, 2, 3, . . ., 10). At each value of x, the value of y (which is just two times x) is calculated, and the values of both x and y are printed. The NEXT X instruction returns the program to line number 10, and increments the value of x by one unit.

The program given above uses the PRINT instruction, which is a very convenient method of obtaining output data from PCs. The output from the above program would appear as follows:

X = 0	Y = 0
X = 1	Y = 2
X = 2	Y = 4
X = 3	Y = 6
X = 4	Y = 8

X = 5	Y = 10
X = 6	Y = 12
X = 7	Y = 14
X = 8	Y = 16
X = 9	Y = 18
X = 10	Y = 20

In line 30 of the program, both X= and Y= are enclosed by quotation marks. Any data enclosed by quotation marks will be printed exactly as shown. Thus in each cycle of program execution, X= and Y= are printed prior to the printing of the values of x and y, respectively. The commas in line 30 direct the printer to allow some spaces between "X=" and the value of x, and "Y=" and the value of y.

The next set of instructions (+, −, *, /, SQR, LOG, SIN, and COS) are mathematical instructions. They represent addition, subtraction, multiplication, division, square root extraction, logarithm computation, sine, and cosine computation, respectively. The instructions are used in equation form. Consider, for example, the following program:

```
10    INPUT U,V,W,X
20    LET G=[(U+V)*(W-X)]/(W*X)
30    LET Y=G+LOG(G)+SIN(G)/COS(G)
40    PRINT "G=",G,"Y=",Y
```

When this program is run, the values of u, v, w, and x are provided via the keyboard. The program then calculates g, which is given by the equation shown in line 20. Line 30 then adds g, its logarithm, and its tangent (since sin(g)/cos(g) = tan(g)) together. Both g and y are then printed.

The ease with which complex mathematical operations can be performed represents one great advantage of the CTL over other types of programming languages.

It should be noted that Boolean algebra can be performed with these mathematical instructions. This is done by replacing AND with *, OR with +, and so on. Chapter 3 should be consulted for the details.

The PEEK command followed by a register number or memory location functions the same as the GET ladder instruction followed by a register number or memory location. The POKE command, followed by a register number or memory location and a byte of data, functions the same as the PUT ladder instruction with a register number or memory location.

GOTO is similar to the JUMP ladder instruction. It directs the program to jump to a specified program line number. It is often used with the IF/THEN pair:

```
80 IF X=72 THEN GOTO 150
```

The GOSUB instruction followed by a program line number and the RETURN instruction are identical to the JUMP-TO-SUBROUTINE and RETURN ladder instructions, respectively.

BASIC, like the ladder diagram, Boolean, and block languages, provides comparison instructions. These instructions are often used with the IF/THEN and GOTO instructions:

```
110 IF X=0 THEN GOTO 200
120 IF X>0 THEN GOTO 300
130 IF X<0 THEN GOTO 140
```

In the program above, if x equals zero, the program skips ahead to line 200. If x is greater than zero, the program skips ahead to line 300. And if x is less then zero, the program proceeds with line 140. (Line 130 in the program above is not necessary and is included only to demonstrate the LESS THAN comparison.)

The BASIC programming language also contains diagnostic instructions. The two shown in Table 9-1 (ERL and ERR) are extremely useful. ERL is used in conjunction with a print statement to indicate the number of the program line that contains an error:

```
110 PRINT ERL
```

ERR is used to identify the type of error that has occurred:

```
290 IF ERR=26 THEN GOTO 500
```

In this example, error number 26 indicates that a FOR instruction has been used without a subsequent NEXT instruction.

REM is a remark line instruction. If a program line is prefaced by REM, the contents of the line are not included as a part of the program. REM instructions can be very helpful to operators:

```
10 REM THIS PROGRAM IS FOR CONTROL OF
20 REM TEMP IN STEAM LINES
30 INPUT TS
40 READ T
50 etc.
```

The END and RUN instructions are necessary to the proper execution of a program. END must be present in all programs. It is the last (i.e.,, highest-numbered) line in the program. The program will not run without the END instruction:

```
990 PRINT "TS=", TS, "T=", T
1000 END
```

When RUN is typed into the keyboard and entered, program execution begins.

Although this section has concentrated on BASIC, there are a wide variety of CTL programming languages. One very interesting CTL, the State Language, is touted by Adatek, Inc. as being clearly superior to all other CTLs. Information of Adatek's State Language can be found in Appendix B.

SUMMARY

There are two principal types of languages used in PC software: low-level, and high-level. The two low-level languages discussed in this chapter were the ladder diagram and Boolean languages. The ladder diagram language is the most popular of the programming languages. It is the oldest PC language, and it is very easy for those who worked with the old relay-logic control systems to understand. Boolean, or mnemonic, language is the second most popular language. Both of these languages are limited in flexibility.

Block diagrams and computer-type languages (CTLs) were the two high-level programming languages discussed in this chapter. Both languages are more powerful and more flexible than low-level languages. Block diagram language is designed to fit the ladder diagram programming format, and thus is popular among those familiar with the old relay-logic control systems. CTLs, such as BASIC, are quite popular among computer-literate operators because they are similar in nature to the languages used to program personal computers. CTLs are the most powerful and flexible of all the PC programming languages.

Chapter 10

PC Applications

HAVING EXAMINED THE BASICS OF PC OPERATIONS, THE HARDWARE, AND THE SOFTWARE associated with PCs, we turn to an examination of PC applications. The application of PCs to industrial situations requires a familiarity with PC programming on the part of the operator. The first section of this chapter deals with the fundamentals of PC programming. The principles set forth in previous chapters will be useful here. Subsequent sections of this chapter will give examples of the utility of PCs in control applications.

ELEMENTS OF PROGRAMMING

GIGO is an acronym for garbage in, garbage out. It is an acronym that should be kept in mind when programming the PC. After all, the PC can do only what it is instructed to do. If the instructions are inadequate or incorrect, the results will be unacceptable.

Before the control program, or *algorithm*, can be written, the programmer must clearly understand the nature of the task to be performed by the PC. It is often useful for the programmer to write out in clear English statements the control task to be performed and the method (e.g., PID) by which the task will be accomplished. For example, in Chapter 5 a control scheme for the control of steam temperature was discussed. The English-language description of this control problem follows:

> The task is to control steam temperature in the line. The method for controlling steam temperature is as follows. Steam temperature will be measured downstream from the water injection site. When the temperature exceeds the set-point value, a pump will be activated. This pump will spray feedwater into the steam line at the water injection site, thereby cooling the steam. The control strategy will therefore be proportional control; i.e., the error signal will be proportional to the difference between actual steam temperature and the set point value.

Once the English-language description of the control problem has been written, the next step is to construct a *flow chart* of the control program. A flow chart is simply a schematic diagram of a program. It employs special symbols to denote the start, end, decision points, and so on of a program. Figure 10-1 shows the most commonly used flow chart symbols. The terminal point symbol indicates the beginning or end of a program. The process symbol is a schematic representation of program instructions that control program processing. The I/O point symbol indicates the presence of instructions that read input data or provide output signals. The decision point symbol indicates instructions in the program that require that a decision be made or that discriminate between states (e.g., equal to, greater than, or less than).

TERMINAL POINT

PROCESS

I/O POINT

DECISION POINT

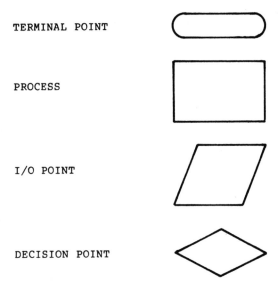

Fig. 10-1. Flow chart symbols.

The flow chart for the steam temperature control program is shown in Fig. 10-2. At the start of the program, the set-point temperature for the steam is entered. This is indicated by the process symbol. The program flows to the I/O point, which reads the actual temperature of the steam in the line. This temperature is then compared with the set-point temperature at the decision point. If the steam temperature in the line is not greater than the set-point temperature, the flow of the program reverts to the I/O point; in other words, the steam temperature is once again read. This series of program steps is repeated until the measured steam temperature is greater than the set-point temperature.

When the steam temperature exceeds the set-point temperature, the program flows from the decision point to another I/O point, which turns the feedwater pump on. This action lowers the steam temperature in the line. Note that the control task and the method of achieving the task are both clearly evident from an examination of the program's flow chart.

Once the flow chart has been drawn, the next step is to write the program. The relay ladder and contact diagrams for steam temperature control given in Figs. 5-8 and 5-10 will not suffice, since those figures show the ladder program for the relay control systems of Fig. 5-6. The PC setup for the control of steam temperature is shown in Fig. 7-2; it employs the ladder program shown in Fig. 10-3. It can be seen that the program is quite simple. The set-point temperature is stored in register number 112, and the

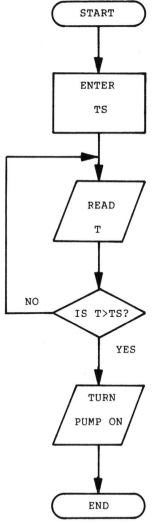

Fig. 10-2. Flow chart for the steam temperature control program.

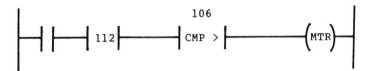

Fig. 10-3. Ladder rung for the control of steam temperature.

output from the thermocouple in register number 106. When the temperature exceeds the set-point value, the CPM > instruction goes ON, and the motor for the pump is activated.

Once you have written the program, you must enter it into the PC. This can be done with the programming devices discussed in Chapter 8 ("Peripherals"). The choice of programming device will be governed by the make and model of the PC being used. The PC's user's manual should be consulted for details.

There are two remaining steps in programming a PC. The program should be *debugged*. This is a sometimes tedious process of locating and correcting mistakes in a program so that the proper control function is performed. This step in the process is made easier if sufficient care is taken in the preceding steps (i.e., the creation of the English description and the flow chart, and the writing of the program). Finally, the program must be *documented*. Documentation involves the preparation of a record of the program and the hardware connections, etc., so that subsequent users of the program will experience little difficulty.

The process presented above has been simplified so that the basic elements of programming are apparent. Many details that are specific to a given PC have been omitted. For example, a specific make and model PC will have certain registers set aside for I/O data and others set aside for general user storage (for information such as set points). These register numbers can be found by consulting the user's manual. The proper register must be addressed by the program, or the program will not perform as desired. Likewise, the input and output voltages that can be handled by different PC I/O terminals can vary. Care should be taken not to assign a 115-volt ac input voltage to a terminal capable of handling only 24 volts ac. A careful reading of the PC's user's manual is usually sufficient to prevent this type of mistake.

APPLICATIONS

As was noted in Chapter 1, there are very few industries today that do not employ programmable controllers. Although most of these businesses are in the manufacturing sector, there are many in the service sector that use PCs. Industries as diverse as aerospace, automotive, bottling and canning, chemicals, entertainment, food and beverage, gas and petroleum, lumber, machining, metals, mining, packaging, petrochemicals, plastics, power, pulp and paper, rubber, and transportation employ PCs in abundance.

This section will provide several examples of how PCs are used in real industrial applications. Before we examine these examples, it will be convenient to introduce a number of standard symbols that will be used in the figures that follow. These are shown in Fig. 10-4. These symbols should be self-explanatory. An example of how these symbols are used is given in Fig. 10-5, which is merely our previous example of steam temperature control. A comparison of Figs. 10-5 and 7-2 will indicate the usefulness of these symbols.

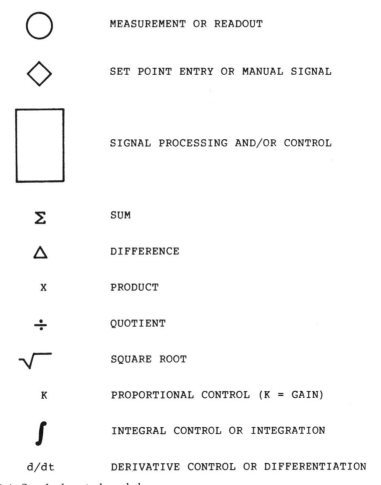

Fig. 10-4. Standard control symbols.

Process Measurement

As the name indicates, programmable controllers find great utility in controlling various processes. Before a process can be controlled, however, key variables in the process must be measured. Sometimes it is not convenient to measure key process variables directly. One example is the mass flow rate of gas. Mass flow rate can be calculated if differential pressure, absolute pressure, and temperature are known. The relationship between mass flow rate and differential pressure, absolute pressure, and absolute

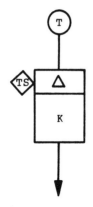

Fig. 10-5. Control of steam temperature.

FEEDWATER PUMP

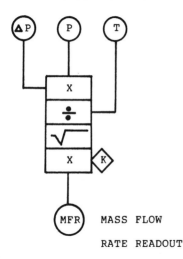

Fig. 10-6. Measurement of mass flow rate.

MASS FLOW

RATE READOUT

temperature is given in Equation 10-1:

$$MASS\ FLOW\ RATE\ =\ K(\Delta P \times P_{absolute}/T_{absolute})^{1/2} \qquad (10\text{-}1)$$

where K is a constant. The measurement of mass flow rate thus requires a differential pressure cell (for ΔP), an absolute pressure transducer, and a temperature transducer calibrated in degrees Kelvin.

A PC setup for the calculation of mass flow rate is shown in Fig. 10-6. Input signals from the ΔP, P, and T transducers are fed into the PC. The ΔP and P signals are multiplied, and this product is divided by T. The square root of this term is then taken, and the whole term is scaled by the factor K.

Another process variable that is calculated is heat transfer rate. Heat transfer rate is a function of flow rate and change in temperature (ΔT), as shown in Equation 10-2:

$$HEAT\ TRANSFER\ RATE\ =\ C \times FLOW\ RATE \times \Delta T \qquad (10\text{-}2)$$

A PC set-up for the calculation of heat transfer rate in a heat exchanger is shown in Fig. 10-7. In the figure, the calculation of mass flow rate is provided by the setup shown in Fig. 10-6. The temperature difference before and after the heat exchanger is taken and is multiplied by the flow rate and the constant C. This yields the heat transfer rate readout.

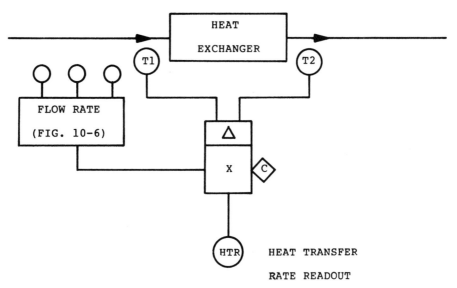

Fig. 10-7. Measurement of heat transfer rate.

Another useful application of PCs in process measurement is the setup shown in Fig. 10-8. This figure shows a sealed reaction vessel in which a chemical reaction is occurring. The temperature in the vessel is controlled, since the reaction becomes explosive at high temperatures. It is desirable to have a measurement of the reaction vessel temperature, the safety margin, and the rate of change of temperature in the vessel.

All of these measurement and control functions are shown in Fig. 10-8. The temperature in the vessel, T, is measured with a thermocouple. This signal is used in performing all four functions. The first function is proportional control of temperature. The control signal from the PC's proportional control program is then supplied to the reactor cooling system. The second function is very simple: the temperature T is displayed on a control panel. The third function is the calculation and display of the *safety margin*. The safety margin is defined as the difference between reaction vessel temperature T and the danger temperature TD, which is the temperature at which the reaction becomes explosive. The final function is the calculation and display of the rate of change of temperature. This is simply the derivative of the temperature signal with respect to time. The portion of the PC program used to calculate the derivative term for PID control can be used for this purpose. An increase in the rate of change of reaction vessel temperature indicates that a safety problem may be developing. Generally, this indication

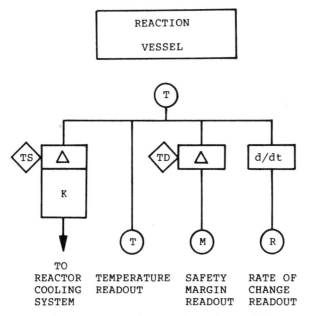

Fig. 10-8. Measurement of temperature, safety margin, and rate of change of temperature.

will occur well before the safety margin or temperature readouts indicate that a problem has developed.

Process Control

It is in the area of process control that PCs are most useful. The examples that follow represent just a few of the many ways that PCs are used in control applications.

Temperature Control. Figure 10-8 presents a setup in which a PC is used to control temperature in a reaction vessel. Figure 10-9 shows a typical reactor and cooling system in greater detail. The chemical reactor is fed by a reactant stream at a specified temperature T. The temperature of the reactant stream is controlled by the valve in the cooling water line. When the valve is opened, cooling water flow to the heat exchanger is increased, and the reactant stream is cooled down. As the valve is closed, cooling water flow to the heat exchanger is decreased, and the reactant stream temperature climbs to the ambient level.

In this uncontrolled (that is, not automatically controlled) system, a human operator would be assigned the responsibility of monitoring the temperature and adjusting the cooling water flow rate accordingly. In other words, the operator becomes the *feedback* mechanism. It is likely that the operator has other responsibilities, so that he is not available for continuous measurement and control.

If it is assumed that the operator measures the temperature and adjusts the cooling water flow rate four times an hour, then Fig. 10-10 could represent the temperature

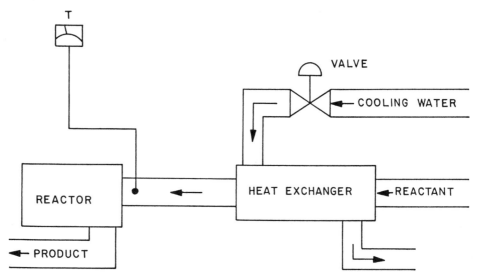

Fig. 10-9. Reactor and cooling system.

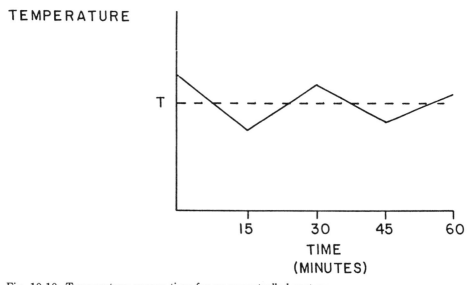

Fig. 10-10. Temperature versus time for an uncontrolled system.

versus time plot for the system over the course of an hour. The figure shows that much of the time the temperature of the reactant stream is very different from the set-point temperature T. The reactor spends most of the hour either producing low yields of product (when the stream temperature is below the set-point temperature) or producing reasonable yields under unsafe conditions (when the stream temperature is above the set-point temperature).

A sudden upset or step change in reactant stream temperature is shown in Fig. 10-11. If, as shown in the figure, the upset occurs shortly after operator has adjusted cooling water flow, the result could be disastrous. The temperature could increase to the explosive limit before the next check is made.

An obvious improvement over this system is the use of a PC programmed with a proportional control algorithm. This is shown in Fig. 10-12. The difference between set-

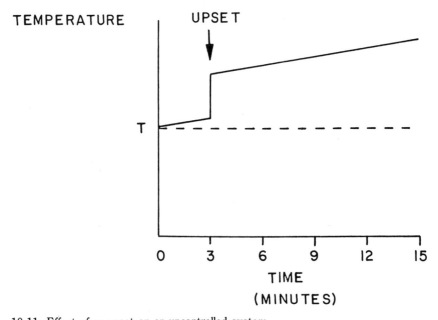

Fig. 10-11. Effect of an upset on an uncontrolled system.

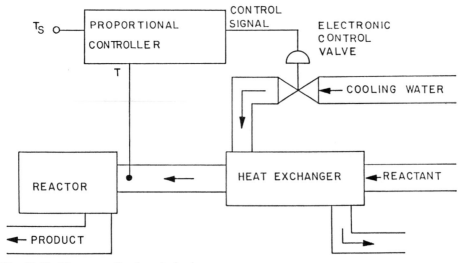

Fig. 10-12. The proportional control scheme.

point temperature and stream temperature is taken, and the difference is multiplied by the gain K. This becomes the control signal that is used to adjust the electronic control valve in the cooling water line.

The proportional control scheme in Fig. 10-12 is shown in Fig. 10-13 with standard control symbols.

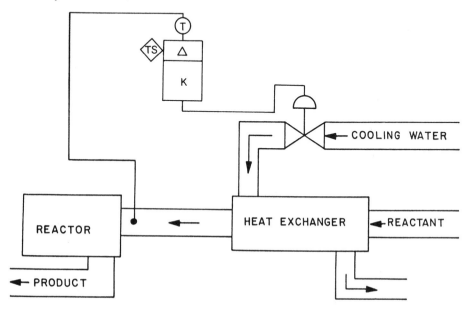

Fig. 10-13. Standard control symbol representation of Fig. 10-12.

Even with the offset that is inherent in proportional control (see Chapter 7), the use of proportional control in this example greatly reduces the amount of time that the reactor spends in the unproductive (low-temperature) or unsafe (high-temperature) regimes.

Consistency Control. Most refineries, chemical manufacturing plants, pharmaceutical plants, and paper mills are criss-crossed with pipes and filled with pumps. The most convenient method of moving raw materials or products back and forth in a plant is by pumping. When the material to be moved is a liquid, there are usually no problems associated with pumps. Generally, however, dry materials cannot be pumped.

If, however, the dry materials are first *slurried* (i.e., mixed with water), then they can be pumped. This is a common method of pumping pigments such as clay. Such a slurry is characterized by its *consistency*, which is just the percentage of dry material in the slurry. A pump that can handle the anticipated consistency is then selected.

Consistency control is important for two reasons. First, if consistency becomes too thick, the pump will not be capable of moving the slurry. Second, the proper mixing of slurries for particular applications will depend upon the availability of a constant concentration of dry material in the slurry. Thus, consistency control is often a critical requirement.

The first step in devising a consistency-control scheme is the selection of a consistency-measuring device. Such a device is shown in Fig. 10-14. The slurry passes through a clear glass or quartz tube through which a light is transmitted. As the consistency of the slurry is increased, the light received by the detector is decreased (that is, the slurry becomes more opaque). When the consistency of the slurry is decreased, the light received by the detector is increased (the slurry becomes more transparent). The detector signal is then treated by the electronics package within the measurement device and boosted to the appropriate level.

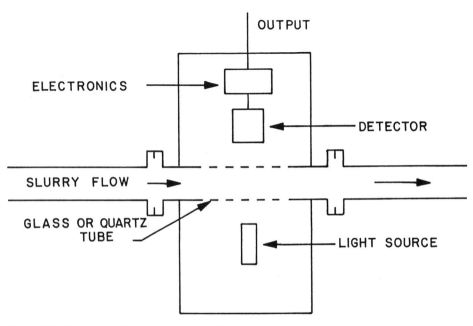

Fig. 10-14. Device for the measurement of consistency.

The consistency measuring device can be calibrated by using slurries of known consistencies that are prepared in a laboratory. In general, the calibration curve would be different for different materials, such as clay, calcium carbonate, wood pulp, etc.

The control scheme for a simple, one-step dilution is shown in Fig. 10-15. This is another example of proportional control. The consistency measuring device is located on the outlet side of the pump. The reason for this is that the pump is the only mixing element in this system, and the slurry should be well mixed before measuring consistency. The consistency signal is then compared with the set-point signal, and the difference is multiplied by the gain K. This control signal is then fed back to the electronic control valve in the dilution water line. If the consistency is too high, dilution water flow is increased. If the consistency is too low, dilution water flow is decreased.

The system in Fig., 10-15 works best when small changes in consistency are required. If large dilution water flow rates are required, consistency control becomes erratic. This can be remedied with the two-step dilution system shown in Fig. 10-16.

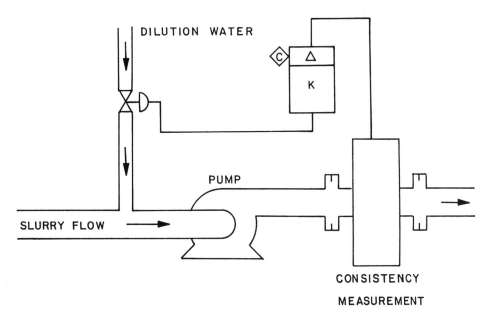

Fig. 10-15. Proportional control scheme for a simple, one-step dilution.

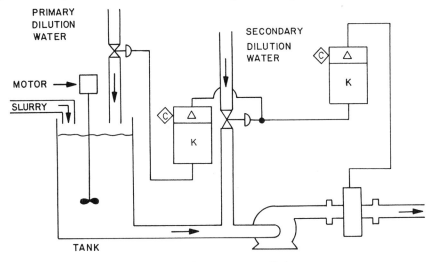

Fig. 10-16. Proportional control scheme for a two-step dilution.

In the two-step system, the secondary dilution control step is just the same as before. What is new is a primary dilution step in which the dilution water is mixed with the slurry in an agitated tank. (The agitated tank is a more efficient mixing element than the pump.) When the consistency error signal is large, such that the secondary control valve is almost fully opened, a second PC proportional control algorithm increases primary dilution water flow. Thus the primary dilution system acts as gross adjustment to consistency, while the secondary dilution system fine-tunes the slurry to the desired consistency.

Level Control. In the previous example, a tank was employed in the primary dilution system. Tanks are used as reservoirs, as mixing elements, and as buffers. In working with tanks, however, care must be taken to assure that they are neither drawn down nor overfilled. The amount of mixture leaving a tank should just equal the amount of liquids entering the tank. This is accomplished very easily with *level control*.

In a level-control scheme, a level indicator provides a signal proportional to the height of the liquid level in the tank. This signal is compared with the set-point signal (desired liquid level), and the difference signal is processed by the PC and fed back to electronic control valves that control the input liquid flow rates to the tank.

There are several methods of indicating liquid level in a tank. We will examine two of the simplest methods. They are the *variable-resistance* method and the *linear-voltage-differential transformer (LVDT)* method.

The variable-resistance method is demonstrated in Fig. 10-17. A vertical rod is attached at one end to a float. The other end of the rod is attached to the wiper of a variable resistor or potentiometer. A potential is applied across the resistor. As the liquid level changes, the position of the wiper changes, as does the voltage at the wiper. Thus, liquid level is transformed into a voltage reading. More accurate variations of this method include the use of the variable resistance in one arm of a Wheatstone bridge, as shown in Fig. 10-18.

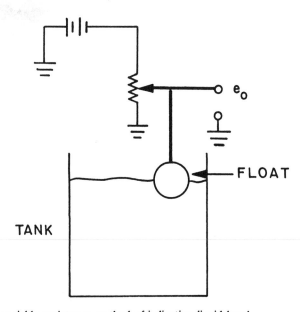

Fig. 10-17. The variable resistance method of indicating liquid level.

The principal drawback to this method is mechanical wear of the variable resistance. After a relatively short operating time, the variable resistor must be changed, since the abrasive action of the wiper tends to destroy the resistor.

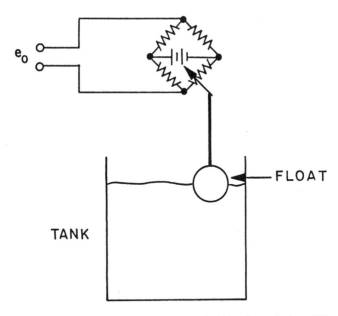

Fig. 10-18. The variable resistance method of indicating liquid level, employing a Wheatstone bridge.

A much better method of determining liquid level (or for that matter, any linear displacement) involves use of the linear-voltage-differential transformer, or LVDT. In fact, the LVDT is the transducer of choice in almost all applications in which position or linear displacement must be measured. This method is shown in Fig. 10-19.

A central coil excited by an ac signal is connected to the vertical float rod. As the level changes, the coil's position between the pair of coils is altered. If the moving coil is centered between the pair, equal voltages are induced in both. The rectified signals from the pair are of equal magnitudes, so that the output signal is zero. If, however, the moving coil is not centered between the pair, voltages of unequal magnitudes are induced in the pair of coils, and the output signal is nonzero (either positive or negative). This signal is used to indicate liquid level (or linear displacement) and is fed to the PC for processing.

The output from an LVDT is not linear over all ranges, as is shown in Fig. 10-20. However, with a little care in the choice of an LVDT, the output can be made linear over the range of interest.

The completed level-control scheme, employing proportional-integral control, is shown in Fig. 10-21. In the figure it is assumed that process demands control the outflow from the tank. In other words, our only degree of freedom in controlling tank level is derived from control of the flow rate of liquid entering the tank. The LVDT provides a signal that is proportional to liquid level, which is compared with the set-point signal. This output is multiplied by the gain K, integrated, and added to itself to provide the proportional-integral control signal. Because of the tank's capacity, oscillations are not a problem, so that derivative action is unnecessary (see Chapter 7).

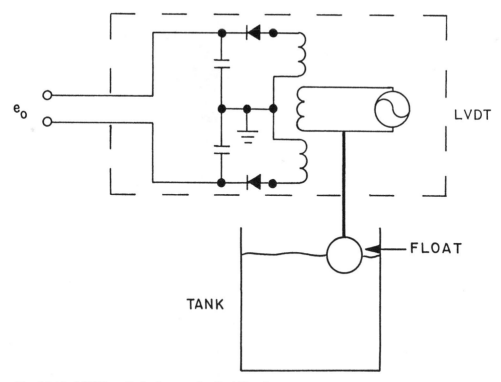

Fig. 10-19. LVDT method of measuring liquid level.

The preceding examples have dealt largely with measurement and control of processes involving liquids. The next example covers new territory: combustion processes.

Fuel-to-Air Ratio Control. Almost all combustion devices, from the internal combustion engine of an automobile to the combustion furnace of an industrial power boiler, require an appropriate fuel-to-air ratio for maximum efficiency. In this example, we consider a process control loop for fuel-to-air ratio. This loop is not the only control loop for the furnace, since the amount of fuel required will be controlled by the demands of the process. That is, the process demand will be the set point for fuel flow, and the set point will vary as process demands change.

Figure 10-22 shows a furnace whose fuel flow rate is controlled by a proportional-integral control algorithm. It is assumed that this is a furnace generating steam for the production of electrical power. As demand for electrical power increases (i.e., as the set point changes), the proportional-integral controller increases fuel flow to the furnace.

In this particular system, only the fuel flow to the furnace changes as demand changes. The air flow remains constant, so that the fuel-to-air ratio departs from the optimum value. The result is a mixture that is either fuel-rich, with incomplete combustion, wasted fuel and increased pollution, or fuel-lean, with insufficient power generation.

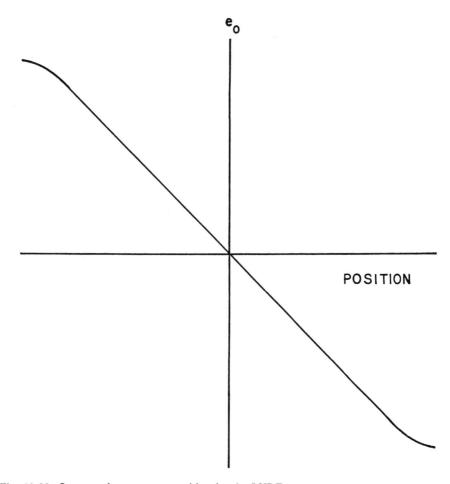

Fig. 10-20. Output voltage versus position for the LVDT.

This situation can be remedied with the control scheme shown in Fig. 10-23. Both fuel and air flow rates are measured using the technique shown in Fig. 10-6 and the ratio is obtained. This ratio is compared with the set-point value. This error signal is integrated and added to itself, and then used to feed an electronic control valve in the air flow line.

This example indicates some of the complexities of process control in industry. Many control loops are used to control one piece of equipment, and the effect of controlling one variable (such as fuel flow) on another (such as fuel-to-air ratio) must be considered.

Control of Steam Pressure. We now examine the control of steam pressure. Steam-pressure control is often critical not only for process efficiency but also for safety. If the pressure limit of a vessel is exceeded, it can explode. For this reason oscillations in steam pressure are not desired, and proportional-integral-derivative control is indicated.

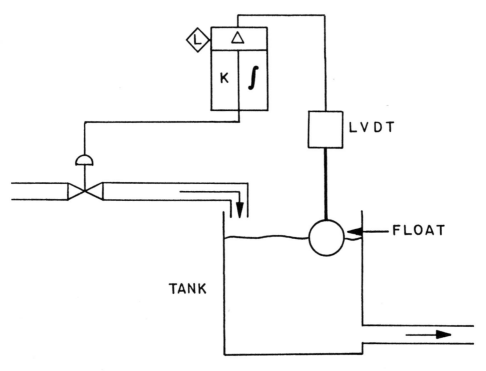

Fig. 10-21. Level control employing proportional-integral control.

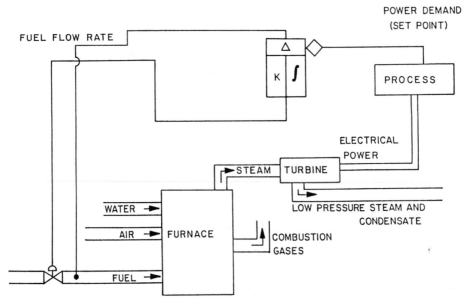

Fig. 10-22. Power boiler proportional-integral control for fuel flow.

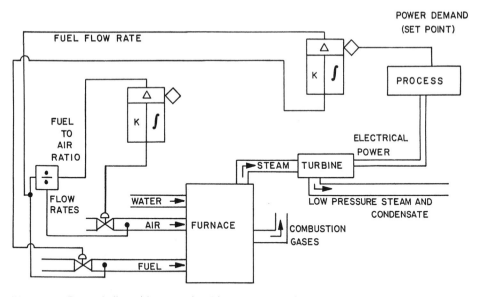

Fig. 10-23. Power boiler with proportional-integral control for fuel flow and proportional-integral control for fuel-to-air ratio control.

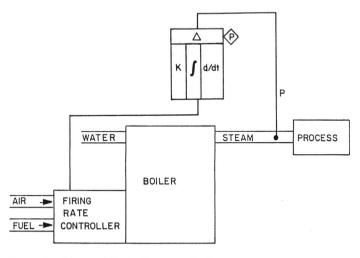

Fig. 10-24. Proportional-integral-derivative control of steam pressure.

A simple steam-pressure control scheme is shown in Fig. 10-24. In this figure, all the considerations of fuel flow, fuel-to-air ratio and ignition signal are represented schematically by the firing rate controller block. This has been done to simplify the figure.

A pressure transducer in the steam line indicates the steam pressure, and the signal from this transducer is compared with the set-point value. The error signal is then subjected to integration and differentiation, and the combined signal (the PID control signal) is then fed to the firing rate controller.

An added complication occurs when process demand is considered. This is illustrated in Fig. 10-25. When the demand for steam changes rapidly, steam pressure tends to fluctuate. In the system shown in the figure, total steam demand will control the fuel flow rate. (Fuel-to-air ratio is controlled in the firing rate controller block.) The steam-pressure feedback loop acts as a trimming control. Multiplication of the two PID signals is appropriate since the amount of correction necessary for a given deviation in steam pressure is proportional to process steam demand (or load). By multiplying the signals, we are using one control signal to compensate equally for both process demand and pressure variations.

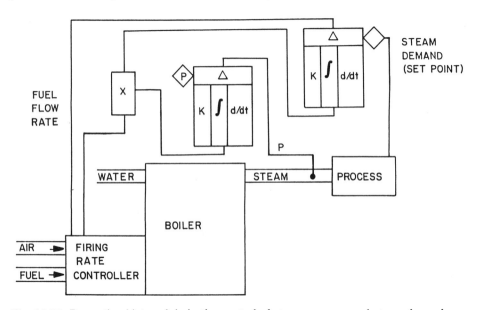

Fig. 10-25. Proportional-integral-derivative control of steam pressure and steam demand.

Most plants use more than one steam or power boiler. This rapidly complicates the control scheme. The general principles outlined above, however, are still utilized in controlling steam pressure and flow.

Effluent Discharge Control. The next example will deal with effluent discharge control. Many process industries use large quantities of water. That is why so many industries are located near lakes or rivers. After the water used in manufacturing is recovered, it is cleaned up and discharged back into the river or lake. Strict regulations are in effect regarding the discharge of process water. The biological oxygen demand (BOD) of the discharged water is limited to certain ranges, the color of the discharged water is monitored, and the effect of discharged water pH and temperature on river (or lake) pH and temperature are strictly controlled.

In this example it will be assumed that all the above effects may be controlled by adjusting the water discharge rate to some optimum value compared with river flow rate. In other words, it will be assumed that there exists a ratio of discharged water volume to river water volume that is an optimum for control of BOD, color, pH, and temperature. The essence of the control scheme will then be control of effluent volumetric flow rate. The feedback signal will be proportional to the river volumetric flow rate.

Three things are needed to determine the volumetric flow rate of the river. They are the river velocity, river level, and river width. This is shown schematically in Fig. 10-26.

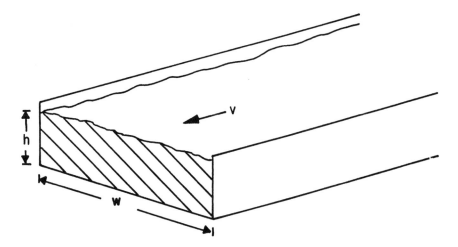

Fig. 10-26. Calculation of river volumetric flow rate.

If it is assumed that the river has a smooth, level bottom of known width, then the height of the river (river level) multiplied by the width will give the cross sectional river area (the shaded portion in Fig. 10-26). Multiplication of this area by river velocity gives the volume of water flowing past the shaded area for a given time. For example, assuming that the river level is three feet, the river width is ten feet, and the river velocity is 60 feet-per-minute, the volumetric flow rate is

$$Q = 3 \text{ ft} \times 10 \text{ ft} \times 60 \text{ ft/min}$$
$$= 1800 \text{ cubic feet per minute} \tag{10-3}$$

Using the conversion factor of 7.48 liquid gallons per cubic foot yields

$$Q = 1800 \times 7.48 = 13,464 \text{ gallons per minute} \tag{10-4}$$

In general, river level and river velocity are not completely independent. That is, a higher river velocity will most likely be found when river level is higher than normal. In order to assure proper calculation of volumetric flow rate, both river level and river velocity should be measured.

In the effluent control system to be presented, a pump will be used to move the effluent from the reservoir to the river. The gallonage moved by the pump can be controlled continuously from zero gallons per minute to the pump maximum value by the pump speed control unit.

The control scheme is shown in Fig. 10-27. Electromagnetic-type flow meters are used to measure river velocity and effluent velocity. Effluent velocity is multiplied by pipe cross-sectional area (previously measured) to obtain effluent volumetric flow rate.

The river velocity signal (from the electromagnetic flow meter) is multiplied by the river level, and this product is multiplied by river width (previously measured). This signal, the river volumetric flow rate, is divided into the effluent volumetric flow rate to yield the ratio of effluent to river volumetric flow rates.

The ratio of effluent volumetric flow rate to river volumetric flow rate is now compared with the optimum value (set-point value). The error signal is subjected to proportional-

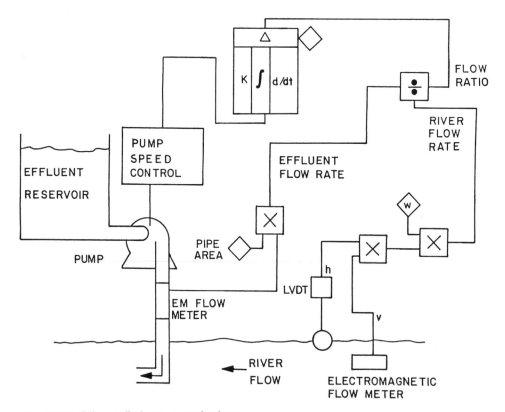

Fig. 10-27. Effluent discharge control scheme.

● Just 3 steps are required to program the Cylinder Movement.

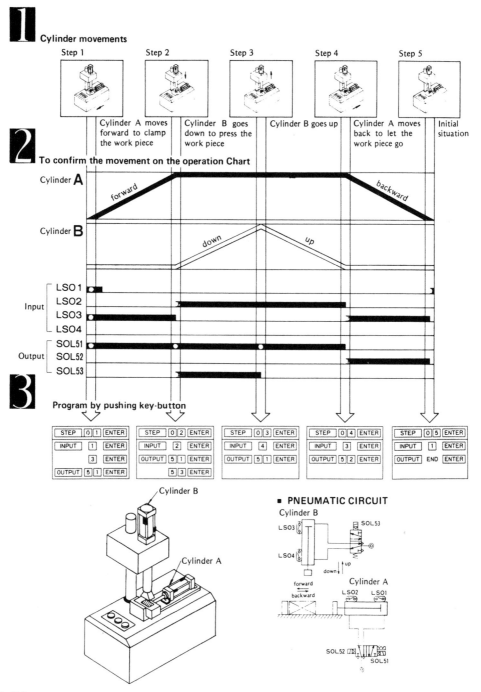

Fig. 10-28. PC control of a pneumatic pressing machine (REPRINTED WITH PERMISSION OF LEHIGH FLUID POWER, INC.).

integral-derivative treatment, and the control signal is fed to the speed control box for the pump. Deviations from the ideal ratio are corrected by adjusting effluent flow rate.

Other Examples

The examples presented previously concentrated primarily on PC control of unit processes that are typical of those found in the chemical industry. PCs are also valuable in controlling other process steps. An example is shown in Fig. 10-28. This figure illustrates the use of a PC to control the operation of a pneumatic pressing machine. This figure is taken from supplier literature for the model TPC-20 programmable controller, which is produced by Miyashita Controls Ltd. of Japan, and marketed in the United States by Lehigh Fluid Power, Inc. (This figure is reproduced with permission of Lehigh Fluid Power, Inc.) In order to facilitate an understanding of the commands

	Name	FUNCTION		Name	FUNCTION
ORDER KEY	STE (step)	New step	MODE SELECT KEY	AUT (auto-)	Remember "END" instruction at end of program
	INP (input)	Input No.		MAN (manual)	Necessary to get manual output function by remote control
	OUT (output)	Output No. to make-on when relative input conditions are satisfied		FWD	Single step operator. Remember "END" at end of program for this mode.
	TMR (timer)	The three digit display shows 0.1 to 99.9 . . (use decimal point key for 0.1 sec. unit) or 1 to 999 sec. OUT → TMR → 9 9 9 → ENT		PRG (program)	Programming. Prohibition remote terminals for accidental and intentional undesired programming (under remote control)
	RPT (repeat)	Set key OUT → RPT → 2 → 02 *Required repetition *The destination step for jumping	AUTO MODE OPERATION KEY	STT (start)	1) Starts program in AUTO 2) Single step operation "STEP" mode
	END	OUT → END → ENT		RTN (CYCLE STOP)	To finish one cycle before the movement is stopped after the button is pushed
INSTRUCTION KEY	ENT	Program writing		(EMERGENCY STOP) STO (stop)	Stops cycle at current STEP & disables outputs
	READ	Read out of entered Program			
	DEL (delete)	To erase the Program			
	INS (insert)	To insert the new program			

Fig. 10-29. TPC-20 instruction set (REPRINTED WITH PERMISSION OF LEHIGH FLUID POWER, INC.)

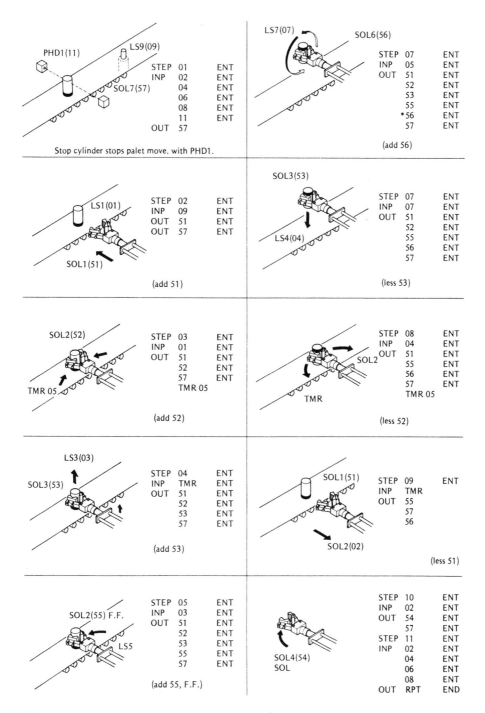

Fig. 10-30. TPC-20 programming for the control of a robot arm (REPRINTED WITH PERMISSION OF LEHIGH FLUID POWER, INC.)

and instructions shown in Fig. 10-28, the instruction set for the TPC-20 is shown in Fig. 10-29.

Programmable controllers are often used in robotics. An example is shown in Fig. 10-30, which is an example of the use of a robotic arm in inverting and removing cylinders from a conveyor belt. This figure is also taken from the literature for the model TPC-20.

Catalogs from suppliers of programmable controllers are very useful in providing typical examples of PC usage. You are advised to consult the supplier literature or to contact the supplier's representative for other potential applications of programmable controllers.

SUMMARY

This chapter has outlined the basic elements of PC programming and has described several examples of PC usage in the process industries. These examples should make clear the great advantage of using programmable controllers for the control of various industrial processes. Additional insights on the use of PCs in process industries can be obtained from PC supplier literature.

Chapter 11
Local Area Networks

THE SUBJECT OF COMMUNICATIONS BETWEEN PCs AND PERIPHERAL DEVICES WAS INTRO-
duced in Chapter 8. Particularly, that chapter dealt with the transmission of data and
control instructions between a PC and a peripheral. The focus of such a communications
system is the complete transmission of data and instructions within one scan of the PC.
Such a network of PCs and peripherals is often called a *control network*.

This chapter also addresses communications, but the emphasis is slightly different
than that of Chapter 8. This chapter will emphasize communications between many PCs,
or between PCs and intelligent devices such as computers, intelligent CRTs, or special
I/O interfaces. In this case the focus is data sharing and system status sharing, not the
communication of I/O data within the time frame of a single scan. This type of network
is often called a *communications network*.

In principle, communication between two PCs (or between a PC and an intelligent
device) is easy to provide. One simply connects the input of the first PC with the output
of the second PC, and the output of the first with the input of the second, using pairs
of wires. This, in fact, is the method that was originally used to provide communications
between PCs.

This method worked very well. The rate of data transmission was limited only by
the baud rates of the two PCs, since only one bit of data was carried by a pair of wires.
The limit of one bit per pair of wires did require that a larger quantity of wire be used,
but the cost was not prohibitive when just two PCs (in relatively close proximity) were
being connected. For two PCs physically separated by a large distance, however, wiring
costs became significant.

With the addition of a third PC, communications became a more complicated affair.
The use of pairs of wire to interconnect all the PCs was prohibited because of the increased
wiring costs. A central computer was then employed to handle communications between
PCs.

In the central computer scheme, three or more PCs are connected to a central computer in a configuration similar to the star configuration of Chapter 8. All communications between PCs are thus routed through the central computer. Although this system minimizes wiring costs, it limits transmission rates to the baud rate of the central computer. Also, if the central computer fails, the entire system goes down.

Given the limitations of the two previous methods of communicating between PCs (wire pairs and central computer), it is understandable why the *local area network* was developed.

A local area network is defined as a system that allows the transmission of data and system statuses between PCs (or between PCs and intelligent devices) at high speeds and over long distances (i.e., two to three miles of cable). A local area network, sometimes known as a *data highway*, can support many PCs or intelligent devices—most local area networks can support at least 100 different PCs or nodes. (Any PC or intelligent device connected to a local area network is called a *node*.) Typically, the communications rate of a local area network is on the order of 57 or 58 kilobaud, or greater.

Local area networks for PCs are quite similar to the business systems local area networks, which are often used to connect office equipment (such as word processors) to a central computer and which allow communications between word processors and computer terminals that are scattered about in various office locations. The two primary differences between business systems local area networks and PC local area networks are that business systems local area networks are slower than PC local area networks, and that business systems networks are not shielded from electromagnetic noise and interference. With some work, business systems local area networks can be adapted for use with PCs.

In addition to the sharing of data and system status, local area networks can be used in *distributed control*. In distributed control, the control functions of one large PC are distributed among several PC subsystems. Currently, distributed control is a very popular and important control concept. An in-depth study of distributed control is beyond the scope of this book. Readers desiring additional information on distributed control are referred to the references in the Bibliography.

There are two major local area network formats. The first format includes those networks that are controlled by a master computer or PC. Communications between PCs in the network are routed through the master computer or PC. This format is known as a *master-slave* local area network (Fig. 11-1). Although this system allows for total control of communications in the network through the use of a master computer, it has the same disadvantage as the central computer scheme discussed previously; i.e., when the master computer (or PC) fails, the entire network goes down. Master-slave networks generally include a back-up master computer (or PC), which is designed to operate the network when the master computer fails.

The second format is one in which there is no master to control the network. Each device, or node in the network, controls its own communications. This format is known as the *peer-to-peer* local area network (Fig. 11-2). Control of such a network is rotated

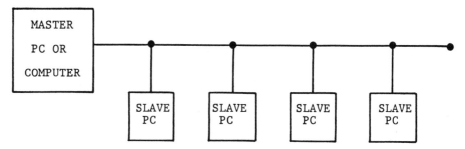

Fig. 11-1. Master-slave format local area network.

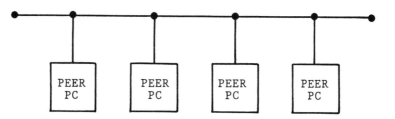

Fig. 11-2. Peer-to-peer format local area network.

among the various nodes in the network. In other words, PC number one has temporary control of the network to effect communications with other network PCs. At some later time, PC number one relinquishes control of the network, and PC number two takes control. After PC number two completes its communications, it passes control of the network to PC number three, and so on. The passing of network control from one peer device to another is know as *token passing*. That is, the PC that has the token has control of the network. The PC in control loses that control when it passes the token to another PC. The peer-to-peer local area network has the advantage that failure of one PC does not bring the system down. On the other hand, the disadvantage is that the schemes necessary to guarantee satisfactory network control are quite complicated, since each peer PC shares control responsibilities equally with other network PCs. This will be discussed later in this chapter.

TRANSMISSION MEDIA

The term *transmission media* refers to the physical components that allow communications between PCs or between PCs and intelligent devices. The two major components of transmission media are the *modem,* or *network adapter module*, and the communications medium, or *bus*.

The term *modem* is an acronym for *MOdulator-DEModulator;* however, the term *network-adapter module* is more descriptive of its function. The modem serves as an interface between the PC or intelligent device and the communications medium (wire, cable, etc.). Essentially, the modem converts the two-level binary voltage from the PC to a two-frequency audio signal for transmission over the wire or cable. Likewise, when the modem receives an audio signal, it converts that signal to a binary voltage for use by the PC. The modem uses a technique known as *frequency-shift keying* (FSK) to accomplish its modulation-demodulation task.

Usually, each PC or node in the network has its own modem. There are, however, some modems that can accommodate several network devices.

The modem is capable of receiving data from a PC at one baud rate and transmitting the data at another, usually higher, baud rate. This is accomplished by including a buffer (or short-term memory, see Chapter 8) in the modem. Data from the PC are accumulated in the buffer and then transmitted at high baud rates over the bus. Received data are stored in the buffer and transmitted to the PC at a lower baud rate.

The term *communications medium* refers to the physical form of the bus. The most common media are twisted-pair conductors, coaxial (coax) and triaxial cables, telephone lines, fiberoptics, and electromagnetic waves (radio and microwave).

Twisted-pair conductors are described by their name. A twisted-pair conductor is composed of two single conductors that are twisted about each other. Until relatively recently, the twisted pair was the most widely used communications medium. It has now been outdistanced by the cables, which will be discussed next. The principal advantage of the twisted-pair conductor is its low cost. When shielded, the twisted-pair conductor is relatively immune to noise and interference. Transmission rates up to 250 kilobaud are attainable with them. The principal disadvantage of the twisted-pair conductor is its nonuniformity. This nonuniformity is manifested in a varying characteristic impedance, which makes impedance-matching virtually impossible. This results in a limited useful range for the twisted-pair conductor (4000 feet or less).

Cable (coaxial and triaxial) represents the most widely used communications medium. Its high degree of uniformity (uniform characteristic impedance) allows its use over large distances (two to three miles). The use of repeaters (devices that receive a signal and automatically retransmit, or repeat, the signal in amplified form) in a network allows the use of cable over very long distances (30 miles or more).

There are two basic types of coaxial (coax) cable: *baseband* and *broadband*. Baseband (or passive) coax allows the transmission of only one signal at a time. Broadband (or cable TV) coax allows the transmission of two or more signals simultaneously on different channels. One broadband channel really consists of two frequencies: a high frequency for transmission and a low frequency for reception. The transmission rate on a broadband channel is usually higher than that for baseband coax (10 megabaud versus 2 megabaud), but the cost of broadband coax can be twice as great as the cost of baseband coax. Broadband coax is generally used when the local area network forms a part of a broadband network. In a broadband network, one channel would be reserved for PC communication while the remaining channels would carry other signals (e.g., video, computer terminals, monitors, etc.).

The use of telephone lines as a communications medium is not an unfamiliar concept to users of computers. A modem designed for use with a telephone handset can be used to provide communications between personal computers all over the country or between intelligent terminals and mainframe computers. The same is true for PC networks. With modems and telephone lines, the PC "local" area network can encompass a very large locality. The cost, of course, is in the form of long-distance telephone bills.

Fiberoptics as a communications medium is in its infancy and is expected to grow rapidly as technological developments occur. There are many reasons for optimism regarding fiberoptics. They are completely immune from electromagnetic noise and interference, smaller and lighter than coax cables, and capable of supporting amazingly high transmission rates (several hundred megabaud) over long distances (five to six miles). The primary losses associated with fiberoptics occur when the bus is tapped (i.e., when a connection is made on the bus to attach a device). Progress in fiberoptics technology is occurring rapidly. One long distance telephone company is, at the time of this writing, in the process of installing lines comprised of 100 percent fiberoptics technology. Widespread use of fiberoptics in PC local area networks should not be far behind.

Electromagnetic waves (radio and microwave) find limited use as a communications medium in PC local area networks. Offshore operations, such as drilling rigs, are the primary users of this medium at present. In theory, the use of microwaves and satellites in geosynchronous earth orbit would allow communications between PCs located a continent apart.

TOPOLOGIES

The word *topology* is used to describe the physical arrangement or configuration of nodes in a network. The daisy chain and star configurations shown in Fig. 8-6 are topologies for multiple peripherals.

The star topology shown in Fig. 11-3 is similar to the star topology shown in Fig. 8-6. The only difference is that Fig. 11-3 shows a star topology involving PCs, not

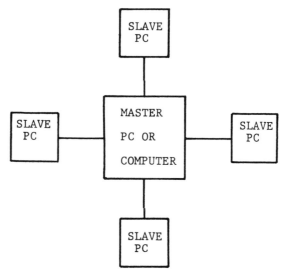

Fig. 11-3. The star topology.

peripherals, with a master PC or computer in the center of the star. Obviously, the star topology works only with the master-slave local area network format. The advantage of the star topology is that communications between each slave and the master can occur whenever needed. The primary disadvantages are that transmission rates are limited by the baud rate of the master, through which all communications must occur, and that failure of the master brings the entire system down. Compared with some of the other topologies, wiring costs can be high.

The daisy chain topology, shown in Fig. 11-4, is used in peer-to-peer format local area networks. A more popular topology, however, is the *common bus* topology, which is shown in Figs. 11-5 and 11-6. Figure 11-5 shows the common bus topology in a peer-to-peer format, while Fig. 11-6 shows the common bus topology in a master-slave format.

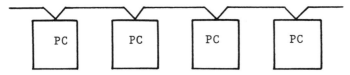

Fig. 11-4. The daisy chain topology.

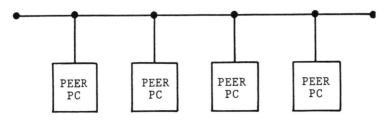

Fig. 11-5. The common bus topology: peer-to-peer format.

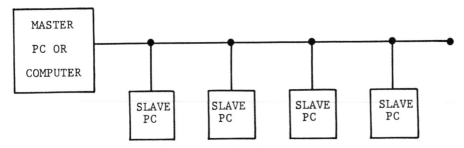

Fig. 11-6. The common bus topology: master-slave format.

The peer-to-peer common bus topology in Fig. 11-5 is sometimes called a *multidrop* topology. The bus itself is a cable that is tapped at each peer PC. The tap used is called a *tee tap*. A short length of cable (drop length of 500 feet or less) is used to connect the main cable (bus) to the modem at each peer PC. The cable that connects the peer PC modem to the bus is called a *drop cable*. When Fig. 11-5 is compared with Fig. 11-4, it becomes apparent that the daisy chain topology is simply a peer-to-peer common bus topology with a zero drop length.

As with other peer-to-peer formats, this common bus topology allows PCs to communicate with each other directly; information is not passed through a master PC. Since no master is present, you need a method of controlling network transactions that allows the PCs to transmit and receive data without interfering with the transmission or reception of another network peer. Methods of controlling transactions on this type of network will be covered later.

The principal advantage of this type of topology is minimized wiring cost. The principal disadvantage is the dependence on a single bus for communication between all PCs in the network. Since a single bus is used, communications between PCs are not as rapid as in other topologies. Also, failure of the bus brings down the network.

The common bus topology shown in Fig. 11-6 is a master-slave format local area network. In this topology, all communications are controlled by the master PC or computer. The master communicates with the slave PCs: any communication between slaves is routed through the master. The master PC addresses, or *polls*, each slave PC for data. When polled, the slave transmits data to the master. The master then has the responsibility of distributing this information to other slaves. This topology shows the disadvantage of other master-slave local area networks: failure of the master brings the system down.

Figure 11-7 shows the *loop* topology. The loop topology is not often used in industrial applications, primarily because the failure of any one PC in the loop is sufficient to bring the entire network down.

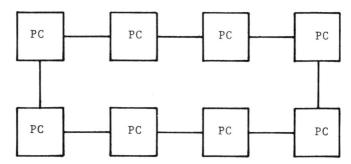

Fig. 11-7. The loop topology.

NETWORK TRANSACTIONS

The two principal methods of controlling network transactions have been mentioned previously. These are *polling* and *token passing*. Polling is used in master-slave format local area networks, while token passing is used in peer-to-peer format local area networks. A third method, *collision detection*, is available but is not often used in local area networks.

The principle behind the polling method of accessing the local area network is simple. The master PC or computer asks a specific slave (polls a specific slave) to transmit data. The master PC or computer then waits a specified period of time. If no data are transmitted by the slave, the master moves on to the next slave in the network. (The master then assumes that the nonresponsive slave is dead, or nonfunctional.) The master continues to poll each slave in the network. Communication between slaves in the network is thus slower than in peer-to-peer format networks, since communication between slaves requires that each slave first be polled and that the master act as an intermediary in sending the message. This type of polling is often called *query-response* communication.

The polling of each individual slave PC in the query-response type of communication is time consuming. The length of time necessary to communicate with the slaves can be shortened by using a variation on the query-response technique. This variation is known as the *broadcast* technique. In a broadcast transaction, the master sends the same message to all the slaves at the same time. Each slave PC receives the message at the same time and carries out the orders or instructions that have been broadcasted. This cuts down on the length of time necessary for the transmission of instructions; however, responses from individual slave PCs still require the polling of slave PCs by the master.

Communication between peer PCs in a peer-to-peer local area network requires a system whereby control of the network is passed from peer to peer. As described previously, token passing is the preferred technique for this purpose. In some respects, token passing is similar to polling. The difference is that each peer in the system takes a turn acting as master of the system. The peer PC which "possesses" the token (i.e., has control of the network) acts as master and polls the other PCs. This can be thought of as *distributed polling*. The PC with the token "passes" the token to the next PC upon completion of its communications. If the next PC does not "accept" the token, the first PC assumes that that particular node is dead, or nonfunctional, and attempts to pass the token to the next PC in the network. This is similar to passing the baton in a relay race: the runner with the baton is the only one allowed to run. The next runner cannot begin until he receives the baton.

Token passing can be time consuming, because each PC is given control of the network for a specified period of time (often called the *token-holding time*).

The PC in control of the network (i.e., the PC with the token) can communicate with another PC in the network by using a *handshaking signal*. The handshaking signal actually consists of two distinct signals. The PC that is to send the message will first send a request-to-transmit signal to the target PC. The target PC, if capable of accepting the message, will then send a clear-to-transmit signal back to the original PC. The transmission of both signals constitutes a handshake. When the handshake is completed, transmission of data between the two PCs is allowed to proceed. The nature of the handshaking signal will vary with the communications protocol used by the network.

Collision detection is a form of network transaction that is more popular in business systems networks than in local area networks. Collision detection works as follows. Each peer PC in the network is equipped with signal-sensing circuitry. This circuitry detects transmissions on the bus and prevents a PC from transmitting when another PC is transmitting. When two PCs begin transmitting at the same time, however, a "collision" of data on the bus is detected. The signal detection circuitry in each PC then disables the PC, so that the transmission of data is interrupted. The PCs then wait a specified period of time, and attempt to retransmit the data. In other words, there is no token passing in this system. A PC can gain control of the network when no other PC is transmitting. Simultaneous transmission by two or more PCs disables the network for a short period of time.

This system works well, provided there are few nodes in the network. Unfortunately, large networks cannot use collision detection since the speed with which data are transmitted is reduced each time a collision is detected, and the probability of collision increases with the number of nodes. That is why collision detection systems are more popular with the (usually smaller) business systems networks.

It should be apparent from this summary of network transactions that the software, or programming, necessary to run the network is fairly complicated. Most PC network suppliers provide software tailor made for their local area networks. The compatibility of a particular make and model PC with the various local area networks should be a major consideration in the selection of a PC system.

PROTOCOL

In Chapter 8, communications standards commonly used for communication between PCs and peripherals were reviewed. These communications standards are often called *protocols*. A protocol is a standard set of rules that specify the format or mode of communication between devices. In this case a device means the same as it did in Chapter 8; a PC, printer, or other peripheral device. The protocol, or communications standard, also includes mechanical aspects (such as wiring) as well as signal levels, timing, and so on.

In Chapter 8, four protocols were mentioned: the RS-232C, the RS-449, the 20mA current loop, and the IEEE 488. In addition to these four protocols, there are two other protocols that work well for local area networks. These are the ISO Open Systems Interconnection Reference Model and the IEEE 802 standard. The physical details of these standard interfaces can be obtained from the International Standards Organization (ISO) and the Institute of Electrical and Electronics Engineers (IEEE), respectively.

The ASCII code (see Chapter 8) remains the most popular protocol for the translation (or encoding) of alphanumeric characters into binary. The ASCII character code is given in Appendix D.

SUMMARY

The local area network provides a convenient method by which communications between PCs are effected. The local area network allows system statuses and data to be displayed on many terminals at one time, and allows for communications between PCs involved in distributed control. This chapter has provided a brief outline of the basic elements of local area networks. Supplier's literature should be consulted for more details.

Chapter 12

Overview of Available Programmable Controllers

THIS CHAPTER PRESENTS TECHNICAL INFORMATION ON PCs THAT WERE AVAILABLE FROM domestic (U.S.) suppliers at the time of this writing. A complete list of domestic PC suppliers, with addresses and telephone numbers, can be found in Appendix A. Further technical information on the PCs listed in this chapter can be obtained directly from the supplier.

Not all of the PCs discussed in this chapter are manufactured in the United States. Some are manufactured abroad and marketed in the U.S.; some are joint ventures between U.S. and foreign manufacturers; and some are manufactured solely in the United States. For this reason, the companies listed in this chapter and in Appendix A are designated as domestic suppliers.

It should be emphasized that the programmable controller market in this country changes very quickly. New PC suppliers spring up each year, and some are eliminated. And of course, many of the older manufacturers remain competitive in the PC market year after year. Given the changes that occur, it is necessary to remind you that the information presented in this chapter is correct only for the time of this writing. The reader interested in staying up to date with current PC manufacturers and systems is advised to subscribe to two trade magazines: *Control Engineering* and *I&CS*.

ASEA INDUSTRIAL SYSTEMS

The ASEA firm, headquartered in Milwaukee, Wisconsin, markets the MasterPiece series of programmable controllers. The series includes the MasterPiece 51, 120, 140, 160, 150T, 170T, 220, 240, 260, and 280. The entire series has PROM and CMOS RAM memories (see Chapter 6). In addition, the MasterPiece 280 also features EEPROM memory. Memory size varies throughout the series. The MasterPiece 51 has 16K of memory, while the Masterpiece 120, 140, 160, 150T, and 170T have 64 K of memory. The MasterPiece 220, 240, 260, and 280 each have 2M of memory. (Note that $1M = 1K \times 1K = 1024 \times 1024 = 1,048,567$ words.) The scan rate per K of memory can be set or defined by the user.

The I/O capability of the MasterPiece series varies from a total of 64 I/O points, or connections, to 5500. The MasterPiece 51 allows only 64 I/O points, all of which are discrete (ON or OFF—see Chapter 7). The MasterPiece 120, 140, 160, 150T, and 170T have 128 I/O points. These can be either discrete or analog. The remaining PCs in the series have 5500 I/O points. Of these, only the MasterPiece 220 is limited to discrete I/O. The others can also handle analog I/O. The MasterPiece 160, 150T, 170T, 220, 240, 260, and 280 PCs can support special interfaces. All of these have ASCII capabilities, and all of these except the MasterPiece 220 are capable of PID control. Communication with peripheral devices is accomplished using the RS-232C serial interface.

None of the PCs in the MasterPiece series can be programmed in ladder language. The only one that uses a low-level language (Boolean) is the MasterPiece 51. The others are programmed with a high-level language. The MasterPiece 51 can be programmed manually, while the others can be programmed with a computer or other intelligent device.

All of the MasterPiece PCs are suitable for use in a local area network. All but the MasterPiece 51 can be used in the master-slave or peer-to-peer formats. The 51 is restricted to the peer-to-peer format.

All of the MasterPiece PCs are capable of performing diagnostic (self-checking) functions. All but the 51, 120, and 220 are capable of performing mathematical functions, such as arithmetic.

ADATEK, INC.

Adatek, Inc., of Sandpoint, Idaho, has been mentioned previously (see Chapter 9 and Appendix B). Adatek produces two PCs: the System 10, and the CMAX. Both PCs have EPROM and CMOS RAM memories. The System 10 has a memory size of 16K, while the CMAX has a memory size of 96K. Adatek lists a scan rate per K of memory of 5 to 20ms for both PCs.

Both the System 10 and CMAX have I/O capabilities of 1272 points, 1176 of which are reserved for discrete I/O and 96 of which are reserved for analog I/O. Both PCs have ASCII and PID capabilities, and both use the RS-232C interface. The System 10 is also capable of using the RS-422 interface.

Neither of the two Adatek PCs uses a low-level language, but both can be programmed with a high-level language. (See Appendix B for a discussion of Adatek's State Language.) Both PCs can be programmed either with a CRT or computer.

Both of Adatek's PCs are suitable for use in a local area network (master-slave or peer-to-peer formats). Both PCs can perform diagnostic checks, and both perform mathematical functions or operations.

ALLEN-BRADLEY CO.

Allen-Bradley, headquartered in Milwaukee, Wisconsin, is one of the two oldest manufacturers of PCs in the United States. The other is Gould, Incorporated. Allen-Bradley produces a wide variety of programmable controllers. These PCs fit into one of two lines: the SLC line or the PLC line.

The SLC line of programmable controllers is a low-end line (small, inexpensive, suitable for simple control functions) and includes the SLC 100 and the SLC 150. Both of these programmable controllers have CMOS RAM and EPROM memories. The CMOS RAM is backed by a battery. This prevents the loss of the contents stored in RAM when power supply to the PC is interrupted. Memory size is small for the SLC line. The SLC 100 can hold 885 words in memory, and the SLC 150 can hold 1200 words. The scan rate for both PCs is 25 milliseconds per 500 words.

Both SLC programmable controllers have an I/O capability of 112 points. All 112 points can be used for discrete I/O, and 24 can be used for analog I/O. The SLC series cannot support special interfaces. Communication with peripheral devices requires the RS-422 interface.

Both the SLC 100 and SLC 150 can be programmed with ladder language. Neither of these programmable controllers can be programmed with a high-level language.

Both SLC PCs are suitable for use in a local area network. Neither PC is capable of performing mathematical functions, but both perform diagnostic functions.

The PLC line of Allen-Bradley programmable controllers includes the PLC-2, PLC-3, and PLC-5 series. The PLC-2 series includes the Mini-PLC-2/05, the Mini-PLC -2/15, the Mini-PLC-2/16, the Mini-PLC-2/17, the PLC-2/20, and the PLC-2/30. All the Mini PLCs except the Mini-PLC-2/15 have RAM and EEPROM memories. The Mini-PLC-2/15 has RAM and EPROM. The PLC-2/20 and PLC-2/30 have RAM memory. Memory size varies throughout the series, as follows: Mini-PLC-2/15—2K; Mini-PLC-2/05 and Mini-PLC-2/16—3K; Mini-PLC-2/17—6K; PLC-2/30—8K; PLC-2/20—16K. Scan rates per K of memory also vary, as follows: Mini-PLC-2/05—15ms; Mini-PLC-2/15—11.5ms; Mini-PLC-2/16 and Mini-PLC-2/17—13ms; PLC-2/20—5ms; PLC-2/30—5ms.

The I/O capability of both the Mini-PLC-2/05 and -2/15 is 128 points, which can be used either for discrete or analog I/O. I/O capability for both the Mini-PLC-2/16 and -2/17 is 256 points for either discrete or analog I/O. The PLC-2/20 and -2/30 are considerably bigger: they both have the capability of 2688 I/O points, all of which may be discrete and 896 of which may be analog. All of the PLC-2 series programmable controllers are capable of supporting special interfaces, such as PID and ASCII. Communication with peripheral devices is accomplished using the RS-232C protocol interface.

All of these PCs can be programmed with ladder language, and all can be programmed with a high-level language. All the PCs in this series can be programmed manually, by CRT, by cassette tape, or by computer.

All of these PCs are suitable for use in a local area network (master-slave or peer-to-peer formats). Allen-Bradley offers its own data highway (or local area network), the

Data Highway II (DH II). This highway adheres to the MAP (*M*anufacturing *A*utomation *P*rotocol) protocol. MAP is viewed by many in the PC industry as the upcoming standard for local area networks.

These PCs can perform diagnostic and mathematical functions.

The Allen-Bradley PLC-3 series includes the PLC-3 and the PLC-3/10. The PLC-3 has 2M of RAM, and the PLC-3/10 has between 16 and 32K of RAM. The scan rate per K of memory for both is 2.5ms.

The PLC-3 has a total I/O capability of 8096 points, all of which may be used for discrete I/O and half of which may be used for analog I/O. The PLC-3/10 has a capability of 2048 total points, all of which may be used either for discrete or analog I/O. Both are capable of ASCII communications and PID control. Communication requires the RS-232C interface.

Both can be programmed with ladder or high-level languages, and both can be programmed manually, by CRT, by cassette tape, or by computer.

The local area network capabilities of these two PCs are the same as for the PLC-2 series listed above. Also, both of these PCs are capable of performing diagnostic and mathematical functions.

The PLC-5/15 programmable controller rounds out the Allen-Bradley line. This PC has RAM and EEPROM memories. Memory size is between 6 and 14K. The scan rate per K of memory is 7ms. It has an I/O capability of 1024 points, all of which may be used for discrete I/O. Only 768 may be used for analog I/O. As with the PLC-3 series, the PLC-5/15 has PID and ASCII capabilities. The RS-232C interface is used in communications.

The PLC-5/15 can be programmed with ladder or high-level languages. As with the PLC-3 series, it can be programmed manually, by CRT, by cassette tape, or by computer. It has the same local area network capabilities as the PLC-2 and PLC-3 series. The PLC-5/15 can perform mathematical and diagnostic functions.

ANDERSON CORNELIUS CO.

Anderson Cornelius Co. of Eden Prairie, Minnesota, markets the MCU-250 and APC-3 programmable controllers. Specifications for the two are similar. Both have PROM and RAM memories. Memory size for both is 80K. The MCU-250 has a scan rate per K of memory of 40ms, and the APC-3 has a scan rate per K of memory of 18ms. Both have total I/O capability of 240 points, usable either for discrete or for analog I/O. Both are capable of PID control. Communication with peripherals requires either the RS-232C or RS-422 protocol interface. Both can be programmed either with ladder or high-level language.

APPLIED SYSTEMS CORP.

Applied Systems Corp. is headquartered in St. Clair Shores, Michigan. It markets three programmable controllers: PC/86, PC/88, and PC/188. All have RAM and EPROM memories, and in addition, the PC/188 has EEPROM memory. Memory size is 256K for the PC/86 and 1M for the PC/88 and PC/188. The scan rates per K of memory are 20ms (PC/86 and PC/88) and 10ms (PC/188).

Both the PC/86 and PC/88 have a total I/O capability of 512 points (512 discrete, 128 analog for the PC/86, 64 analog for the PC/88). The PC/188 has a total I/O capability of 1000 points (1000 discrete, 128 analog). All are capable of PID control. Both the PC/88 and PC/188 use the RS-232C, RS-422, or RS-488 protocols. The PC/86 uses the RS-232C or RS-422 protocols for interfacing with peripheral devices.

All three PCs can be programmed with ladder or high-level programming languages, and all are suitable for use in local area networks.

AUTOMATION SYSTEMS INC.

Eldridge, Iowa is the home of Automation Systems Inc., which produces the PAC-5 series of PCs. The four PCs in this series are the PAC-5 5100, the PAC-5 5110, the PAC-5 5200, and the PAC-5 5300. All four have RAM, EPROM, and EAROM memories. The 5100, 5110, and 5200 have between 8 and 64K of memory, and the 5300 has 64K of memory. The scan rate per K of memory is 1.1ms.

The total I/O capability of the series varies with model number. The 5100 has 64 total I/O points, the 5110 has 128, the 5200 has 384, and the 5300 has 1024. These can be used either for discrete or analog I/O. All four are capable of PID control, and all four use the RS-232C or RS-422 interfaces.

All four can be programmed in ladder or high-level language. The series can be programmed with a computer. All the PCs in the series can be used in local area networks (master-slave or peer-to-peer). All four can perform diagnostic and mathematical functions.

BAILEY CONTROLS CO.

Bailey Controls of Wickliffe, Ohio is a subsidiary of Babcock & Wilcox (a McDermott Company). Bailey Controls features the PC-90 programmable controller. The PC-90 has RAM, EEPROM, and UV-EPROM memories. Memory size is 256K. The scan rate per K of memory is 2ms.

The PC-90 has a total I/O capability of 1000 points, which can be either discrete or analog. It has PID and ASCII capabilities. PID control is performed via analog calculations. The communication protocol is the RS-232C. The PC-90 can be programmed either manually or with a computer. It uses relay ladder, functional block diagrams, or high-level language (such as BASIC and C).

The PC-90 can be used in master-slave or peer-to-peer communications with the Bailey Local Area Network or Plant Loop. It is also useful in distributed control (using the NETWORK 90 Distributed Control System). In addition, up to 64 sites can be tied to the PC-90 via remote I/O.

The PC-90 can perform diagnostic and mathematical functions, such as arithmetic, matrix manipulation, function generation, timing, counting, and pulse generation.

BARBER-COLMAN CO.

Barber-Colman (Loves Park, Illinois) markets the EDAC programmable controller. The EDAC was not designed as a general purpose PC; rather, it was designed especially for the heat processing industries. It is generally used in such heat processes as sintering, carburizing, tempering, brazing, or annealing. It contains up to 40K of UV-PROM memory. The scan rate per K of memory is 2ms.

I/O points are contained in racks of 128 I/O points per rack. A maximum of three racks can be used, yielding 384 total I/O points. All of these can be used for discrete I/O, but only 84 can be used for analog I/O. The EDAC is capable of PID control and is compatible with ASCII interfaces. Communication between the EDAC and a host computer uses the RS-232C interface.

The EDAC is programmed with ladder language, using a CRT. It is compatible as a slave in a local area network that contains a host computer. It performs diagnostic functions.

CINCINNATI MILACRON CO.

Cincinnati Milacron of Lebanon, Ohio, markets three PCs: the APC105, APC500 Relay, and APC500 MCL. The APC105 features 1K of CMOS EEPROM memory, while the two APC500s feature 32K of CMOS RAM memory. The scan rate per K of memory for the APC105 and APC500 Relay is 5ms. For the APC500 MCL, the scan rate is 4.5ms.

The APC105 has a total of 64 I/O points, all of which are reserved for discrete I/O. The APC500 Relay has a total of 512 I/O points, all of which may be used for discrete I/O and 64 of which may be used for analog I/O. The APC500 MCL has a total of 2048 I/O points. All of these may be used for discrete I/O; however, only 128 may be used for analog I/O. The two APC500s are capable of PID control and are ASCII compatible. These two use the RS-232C or RS-422 interface.

The APC105 and APC500 Relay PCs can be programmed only in ladder language using either a CRT or computer, while the APC500 MCL uses only high-level language. It too can be programmed by CRT or computer.

All three of these PCs are compatible with remote I/O interfaces, host computers, and peer-to-peer local area networks. The APC500s perform diagnostic and mathematical functions.

CONTROL TECHNOLOGY CORPORATION

Control Technology, headquartered in Hopkinton, Massachusetts, markets seven different programmable controllers. These are the 2008, 2008R, 2021, 2400, 2400iE, 2800, and 2800iE. The first three in this list have EEPROM and RAM memories, with memory size of 2K. The 2400 and 2800 have 4K of EEPROM memory, while the 2400iE and 2800iE contain 128K of RAM and NOVRAM. Scan rates per K of memory are: 1.8ms for the 2008, 2008R, and 2021; 1.8ms for the 2800; and 0.9ms for the 2400iE and 2800iE.

The Control Technology PCs span the range of I/O capabilities. The 2008 and 2008R PCs support a total of 28 I/O points, 20 of which are reserved for discrete I/O and 8 for analog. The 2021 has a total of 44 discrete I/O points. The 2400 has a total of 64 I/O points (64 discrete and 32 analog). The 2800 has a total of 384 I/O points (256 discrete and 128 analog). Both the 2400iE and 2800iE can accommodate a large number of I/O points: 10K (8 K discrete and 4K analog). The 2400iE, 2800, and 2800iE are capable of PID control.

Interface protocols also vary, depending upon the model. The 2008, 2008R, and 2021 use parallel interfaces. The 2400 uses the RS-232C. Both the 2400iE and 2800iE use either the RS-232C or RS-422 protocols. The interface protocol for the 2800 is proprietary.

None of these PCs can be programmed with ladder language. The 2400 can be programmed with Boolean language, and the others use a high-level language.

CROUZET CONTROLS, INC.

Crouzet Controls, Inc. of Schaumburg, Illinois markets three French-made programmable controllers: the CMP-31, CMP-34, and CMP-340. The CMP-31 offers 4K of EPROM memory, while the CMP-34 and CMP-340 offer 8K of EPROM memory. The scan rate per K of memory is 7ms.

The CMP-31 and CMP-34 have a total system I/O capability of 32 points. All 32 can be used for discrete I/O, and 16 may be used for analog I/O. The CMP-340 has 64 discrete I/O points. None of these PCs offer PID control capabilities. The CMP-34 and CMP-340 communicate using the RS-232C serial interface.

All three PCs are programmed with a high-level language called GRAFCET (functional flow chart language). None can be programmed with ladder language. All programming of Crouzet PCs is done manually.

The CMP-31 cannot be used in local area networks. The other two can be used with a host computer or in peer-to-peer format networks. All three Crouzet PCs can perform diagnostic functions.

DIVELBISS CORPORATION

The Divelbiss Corporation of Fredericktown, Ohio, markets the Baby Bear Bones and Bear Bones programmable controllers. Like the SLC 100 and 150 (Allen-Bradley), these PCs are relatively low in cost, have fewer I/Os and features than more expensive models, but can be practical alternatives to more expensive PCs when complicated control schemes are unnecessary. Both PCs have 4K of EPROM memory with scan rates per K of memory of 5ms.

The Baby Bear Bones sports a total I/O capability of 26 points, all of which may be used for discrete I/O and 25 of which may be used for analog I/O. The Bear Bones has a total I/O capability of 250 points. All of these may be used for discrete I/O, and 25 may be used for analog. Neither PC offers PID control capabilities. Neither have local area network capabilities.

Both the Baby Bear Bones and the Bear Bones are programmed with ladder language. Both can be programmed either with a computer or with a Divelbiss ICM-PR-05 programming device. The programming device can be used to burn programs on to the EPROM for on-line operation.

DYNAGE CONTROLS

Dynage Controls of Cromwell, Connecticut, offers the SAFE 8000 programmable controller. It offers 40K of CMOS RAM memory.

Total I/O capability is 2176 points (2048 discrete, and 128 analog). The SAFE 8000 is capable of PID control, and communication is accomplished with the RS-232C or RS-422 serial interfaces. It uses a high-level programming language.

EAGLE SIGNAL CONTROLS DIV.

Eagle Signal Controls, an Austin, Texas, division of Wickes Manufacturing Company, distributes the Eptak series of programmable controllers. The series includes the Eptak 100, 120, 225, 245, and 7000. The Eptak 100 has 250 statements worth of EEPROM, while the Eptak 120 allows 520 statements of RAM and UV-PROM. The remaining PCs (Eptak 225, 245, 7000) have battery-backed CMOS RAM and vary in memory size from 8 to 16K (Eptak 225 and 245) to 48K (Eptak 7000). Scan rates for the Eptak 100 and 120 are 20ms, while the Eptak 225, 245, and 7000 have scan rates per K of memory of 39ms, 46ms, and 1.5ms, respectively.

The Eptak 100 has the smallest I/O capability—16 points all of which may be used for discrete I/O. The Eptak 120 has a total of 66 I/O points (all discrete), while the Eptak 225 has 128 I/O points (all discrete). The Eptak 245 also has 128 I/O points, all of which may be used for discrete I/O and 32 of which may be used for analog I/O. The Eptak 7000 has a total of 2048 I/O points, all of which may be used for discrete I/O, and 840 of which may be used for analog I/O. Only the Eptak 245 and 7000 are capable of PID control. Only the Eptak 225, 245, and 7000 are suitable for use in a local area network. The communication protocol for these three PCs is either the RS-232C or the RS-422 serial interfaces.

All of these PCs can be programmed with relay ladder or Boolean languages. Programming can be performed manually, with a CRT or computer.

The Eptak 7000 is suitable for use with remote I/Os, and the 225, 245, and 7000 are suitable for use either with a host computer or in peer-to-peer networks. These three PCs also perform mathematical and diagnostic functions.

EATON CORPORATION

The Cutler-Hammer Products Division of Eaton Corporation of Milwaukee, Wisconsin, markets three programmable controllers: the MPC1, the D100, and the D500.

The MPC1 is a modular programmable controller for applications up to 128 I/O. The MPC1 comes in discrete and analog versions. The discrete version is the MPC1C10.

This supports only 32 I/O points. The expandable PC is the MPC1C20. It supports the basic 32 I/O points plus up to three expansion chassis of 32 I/O points each (a total of 128). The discrete MPC1 is programmed with a hand-held programmer. The basic analog programmable controller (no expansion chassis) is the MPC1C12. This supports 32 I/O points only. The MPC1C22 has the basic 32 I/O points and can support up to three I/O expansion chassis of 32 I/O points each, yielding 128 total I/O points. The analog MPC1 is programmed with a hand-held programmer. Neither version is capable of PID control.

The MPC1 sports 2K of RAM and EEPROM memories. The scan rate per K of memory is 9ms. The MPC1 programmer has a battery-backed RAM memory and can be powered by the MPC1 or an optional desk top power supply. It can write programs to the PC, EEPROM chips, or a printer. All programming of the MPC1 is done with ladder language.

The MPC1 can be used with remote I/O, and performs basic mathematical functions (addition, subtraction, greater than, less than, equal to, etc.). The MPC1 also performs diagnostic functions. It is not well suited for use in a local area network.

The D100 PC, like the Allen-Bradley SLC 100 and Eagle Signal Controls Eptak 225 and 245, is a small, low-cost programmable controller. The D100 is actually a family of products for use in a wide range of applications. The basic unit is offered in three different sizes: 20 I/O, 40 I/O, and 40 I/O high performance. With expansion units, the PC is expandable to 120 I/O (all of which are reserved for discrete I/O). Like the MPC1, the D100 does not offer PID control capability. The D100 is programmed by a hand-held programmer (D100PG10).

The D100 (40 I/O and 40 I/O high performance) has 1K of memory, while the smaller version (20 I/O) has ½K of memory. Memories are RAM and PROM. Scan rate per K of memory is 5ms.

As with the MPC1, the D100 is programmed in ladder language. In addition to being programmed by the programmer, the D100 can be programmed with a CRT or computer. It features diagnostic capabilities.

The D500 is the larger, full-function PC. It offers all the capabilities of large frame PCs, but in a compact, economical design. There are two D500 processors: the CPU25 and the CPU50. The CPU25 features 256 I/O and 4K of memory, while the CPU50 offers 512 I/O and 8K of memory. Memories are RAM and EEPROM. Scan rate per K of memory is 1ms.

The D500 PCs use ladder language. The instruction set is fairly extensive. It includes timers, counters, MCR, and JUMP functions. Additionally, the D500 PCs perform mathematical functions, such as arithmetic, compare, square root, and trigonometric functions. The D500 is capable of PID control and performs diagnostic functions.

The D500 can be programmed with the D500GPP40 Graphic Program Panel, CRTs, cassette tape, and computers. It is suitable for use with remote I/O, and in local area networks. Eaton/Cutler-Hammer offers the EasyNet Data Highway System in a peer-to-peer format. The network interface is the RS-422 serial protocol.

ELECTROMATIC CONTROLS CORP.

Electromatic Controls Corp. of Hoffman Estates, Illinois, markets a broad range of PCs manufactured in Denmark. All of the Electromatic Controls PCs display 758 steps of CMOS RAM memory, all with a scan rate per K of memory of 30ms.

The PLC-30-0606 is a small PC, offering 12 I/O. It is programmed with ladder language. The PLCF-22-1616 and PLCF-32-1616 are slightly larger. They offer 32 total I/O points (all discrete), and programming in the Boolean language. In addition, they offer remote I/O capabilities.

The PLC-23-0816 and PLC-33-0816 offer 56 discrete I/O points, and both ladder and Boolean programming languages. These PCs can be programmed manually or with cassette tape.

The two biggest Electromatic Controls PCs (in terms of I/O) are the PLCF-22-3232 and the PLCF-32-3232. These offer 64 total I/O, all of which may be used for discrete I/O and 8 of which may be used for analog I/O. Ladder and Boolean languages can be used with these two PCs. The PCs can be programmed manually or by cassette tape.

None of the Electromatic Controls Corp. PCs are suitable for use in local area networks.

ENCODER PRODUCTS CO.

Encoder Products of Sandpoint, Idaho offers two PCs: the 7152 and 7252. Both feature battery-backed RAM and EEPROM memories. For both, total memory size is 96K, with 64K of application memory available. The scan rate per K of memory depends upon program structure.

Both feature a total of 408 I/O points (408 discrete and 238 analog). Both can be used with remote I/O as well as PID and ASCII I/O. With the use of repeaters, communication with remote I/O can be maintained up to 40,000 feet. Communication is accomplished through the use of the RS-232C or RS-422 serial interfaces.

Both the 7152 and 7252 can be programmed with the BASIC language (high-level language). Neither can be programmed in ladder language. Both can be programmed using a hand-held programmer, CRT, or personal computer. Both PCs perform standard arithmetical functions, trigonometric functions, and matrix manipulations. They both perform internal diagnostic functions.

ENTERTRON INDUSTRIES

Entertron Industries of Gasport, New York offers three PCs: the SK 1600, the SK 1600R, and the SK 1800. All three feature EPROM memory. The SK 1600 has 2K, the SK 1600R has 4K, and the SK 1800 has 8K. Scan rate per K of memory for the SK 1600 and SK 1600R varies between 15 and 30ms. For the SK 1800, the scan rate varies between 0.8 and 1.6ms per K of memory.

Total I/O capabilities for the SK 1600, SK 1600R, and SK 1800 are 64, 56, and 88 points, respectively. These are all restricted to discrete I/O.

All three can be programmed with ladder language using a computer. All three perform basic mathematical functions.

THE FOXBORO CO.

Foxboro of Foxboro, Massachusetts, produces two PCs: the 3PC-3A, and the 3PC-4A. Both feature EEPROM and RAM memories. The 3A has a total of 32K of memory, while the 4A has 48K of memory. Both feature scan rates per K of memory of 1ms.

The 3A has a total I/O capability of 512 points, and the 4A has a total capability of 2048 points. In both, all I/O can be used for discrete I/O, but only 128 points can be used for analog I/O. Both are capable of PID control. Communication is accomplished using the RS-485 interface.

The 3PC-3A uses ladder language, while the 3PC-4A uses a high-level language. Both can be programmed with a CRT or computer. Both feature remote I/O capabilities. Both are suitable for use in a peer-to-peer local area network. Foxboro provides a data highway that satisfies the MAP protocol. Both PCs are capable of performing basic mathematical functions.

FURNAS ELECTRIC CO.

Furnas Electric Co. of Batavia, Illinois, offers two major lines of programmable controllers: The microPC/96, and the PC/96.

The microPC/96 line of PCs is one of the exciting, low-end lines of PCs that are relatively inexpensive yet offer a wide variety of control capabilities. These are modular PCs that come in very compact packages. Power consumption is low, and a wide variety of peripheral devices are available. The main assembly for the microPC/96 consists of power supply, memory, processor, and I/O. There are three basic models available: a 12 I/O model, expandable to 32 I/O; a 20 I/O model, expandable to 40 I/O; and a 40 I/O model, expandable to 80 I/O. These are all programmed with both the relay ladder language and a Boolean instruction set that includes relays, timers and counters, and conditional jumps. The programmable memory is CMOS RAM. The 12- and 20-I/O models include 320 steps of memory, while the 40 I/O model includes 890 steps of programmable CMOS RAM. Additionally, the 20- and 40-I/O models may use optional EPROM memory. The CMOS RAM is backed up by a lithium battery with a five year life. Typical scan rates for the 12- and 20-I/O models are 3 to 13ms, while the 40-I/O offers a typical scan rate of 10 to 30ms. All the I/O for the microPC/96 are reserved for discrete I/O.

The microPC/96 performs diagnostic functions (low battery and process-error indication).

In addition to I/O expansion assemblies, the microPC/96 is compatible with other, special-purpose expansion assemblies. For example, a positioning counter assembly that provides 400 set points of motor position control and positioning operations is available. Also available is a timer expansion assembly that offers a variable time range (selectable from 0.1s to 600s per point). A high-speed counter adapter is also available. This accepts external high-speed inputs up to 1 kHz. Peer-to-peer communications between PCs can be effected with a wire communication link assembly. This exchanges up to 32 points of information for each PC over a maximum distance of ten meters. The microPC/96 is also compatible with optical fiber cables, through the use of an optical fiber communication link. This too exchanges up to 32 points of information for each PC, but over a distance of 50 meters.

The microPC/96 programmable controllers can be programmed manually, by cassette tape, or by computer.

The PC/96 is larger than the microPC/96. It contains 2K of CMOS RAM user program memory (16 bit words). A typical scan rate per K of memory is 20ms. It includes a total 256 I/O, all of which may be used for discrete I/O, and 56 of which may be used for analog I/O. The PC/96 has PID control capabilities.

The PC/96 uses the LOLa™ programming language, which is a proprietary ladder language with a fairly extensive instruction set. The instruction set includes relays, latch and unlatch, counters, arithmetical functions and value comparison, move, skip, and MCR.

The PC/96 can be programmed with a computer. It is fully compatible with IBM devices. It is suitable for use in peer-to-peer local area networks and in networks that are mastered by a computer. System Interaction Modules (SIMs) are used to display or transmit data and to perform diagnostic functions. The PC/96 is MAP compatible.

GEC AUTOMATION PROJECTS

GEC Automation Projects of Southfield, Michigan distributes four lines of U.S./U.K. PCs. These lines are the Microgem, the Minigem, the GEM-80/141 and 142, and the GEM-80/300 series.

The Microgem and Minigem are low-end PCs. Both have battery-backed CMOS RAM memory. The Minigem contains 2K of memory, while the Microgem contains 4K of memory. Scan rate per K of memory is 5ms.

I/O capabilities for the Microgem and Minigem are 24 and 64, respectively. Microgem I/O is restricted to discrete I/O, while the Minigem can support a maximum of 10 analog I/O. Communication between the PC and peripherals employs the 20mA current loop interface protocol for both PCs.

All the GEC Automation Projects PCs are programmed with ladder language. The Microgem is programmed manually using the Microgem Programmer. The Minigem may be programmed by CRT or computer. Both can be used in peer-to-peer local area networks, and the Minigem can be used with remote I/O and a host computer. Both perform diagnostic functions. The Minigem is capable of performing mathematical functions.

The GEM-80/141 and 142 are much larger. They have 128K of user memory that is battery-backed CMOS RAM. Total memory size (including EPROM) is 256K. Scan rate per K of memory is 1.25ms.

The GEM-80/141 has 512 I/O points, all of which may be used for discrete I/O and 32 of which may be used for analog I/O. The GEM-80/142 has 2048 I/O points, all of which may be used for discrete I/O and 128 of which may be used for analog I/O. These two PCs are fully compatible with the GEM-80 family, which includes the 300 series. Communication involves the use of the RS-422 interface. The GEM-80/141 and 142 are capable of PID control. These PCs can be programmed with a CRT or computer. Network capabilities include remote I/O, host computer, and peer-to-peer communications. Both perform mathematical and diagnostic functions.

The GEM-80/300 series includes the 301, 302, 303, 311, 312, and 313 programmable controllers. The series includes powerful PCs that can handle up to 8192 I/O points.

The 300 series PCs come with four memory options: 192K, 256K, 384K, or 512K. Memory may be expanded up to approximately 1.5M. The expansion memory module carries both RAM and EPROM. Scan rate per K of memory is on the order of 1.25ms.

The I/O capability, as previously noted, extends to 8192 I/O, all of which may be used for discrete I/O and 256 of which may be used for analog I/O. Communication requires one of four serial interface ports provided on the PC. These include the RS-422 and 20mA current loop protocols. The 300 series is suitable for use in local area networks, including the STARNET or CORONET local area networks.

As noted previously, all of the GEC Automation Projects PCs use ladder language. The 300 series includes a comprehensive instruction set (relay, arithmetical, signal processing, PID, BCD conversion, etc.). The GEM-80/311, 312, and 313 include additional instructions, such as PIDCON, TCONST, and RAMPGEN. These are all special programming functions. Program capacity depends upon memory size. The 192K memory can accommodate approximately 5000 user instructions. The 256K memory can handle about 12,000 instructions. The 384K and 512K memories can handle approximately 24,000 and 36,000 instructions, respectively. Instructions can be processed at the rate of approximately 5ms per 1000 instructions.

GE FANUC AUTOMATION

The GE Fanuc Automation company of Charlottesville, Virginia is a joint venture of General Electric and Fanuc Ltd. of Japan. It offers a broad range of programmable controllers that can be classified by series numbers. These are the Series One™, Series Three™, and Series Six™ PCs.

The Series One PC is a compact PC for use as relay replacer. It is so small that it easily fits into a shoebox. Its memory size ranges up to 1724 16-bit words. Memory availability can be specified, as follows: CMOS RAM—700 words; CMOS RAM—1724 words; EPROM—1724 words. The scan rate depends upon the number of words. A scan rate of 12ms is available for 250 words, 20ms for 500 words, 40ms for 100 words, and 60ms for 1700 words. The CMOS RAM is battery-backed (a lithium battery with a two year life).

The Series One offers a basic 64 I/O points, which can be expanded to 112 I/O points by adding two additional racks. These are discrete I/O only. It supports remote I/O, with transmission rates of 19.2 Kbaud over 3000 feet. Remote I/O can be added at up to three locations, with 32 I/O points maximum at each location. Communication with remote I/O uses twisted pair wires. Data communication uses the RS-422 serial interface, also at a rate of 19.2 Kbaud.

The Series One PC is programmed with a relay-ladder based Boolean language. The instruction set includes relays, latch and unlatch, timers, counters, and MCR. Programming is accomplished with a hand-held programmer or with a Workmaster™ Programmable Control Information Center. The Workmaster™ is a portable, industrial-grade IBM Personal Computer. Programs can also be loaded from a cassette tape player.

The Series One works well in a local area network as a slave PC. GE Fanuc offers the GEnet Factory Local Area Network, which uses a broadband bus and token passing. GEnet complies with the MAP protocol and uses the RS-422 serial interface.

The Series One Junior PC is a low cost, compact version of the Series One. It is designed for control applications involving four to 60 relays. The Series One Junior is contained in one package that includes 24 I/O points, 700 words of memory, and a built in 2 kHz high speed counter. The 700 words of CMOS memory can be replaced by 700 words of EPROM, if desired. I/O capacity for the Series One Junior can be expanded to a total of 64 I/O points by using the standard Series One rack and discrete I/O modules. The Series One Junior PC uses the same programming language, programmers, cassette tape recorder, printer interface and data communications as the Series One PC.

The Series One Plus programmable controller is an upgraded version of the Series One. It is still small enough to fit into a shoebox. It offers a total of 168 I/O points. Its memory is the same as the Series One. Scan rates, however, are faster (8ms/500 words, 12ms/1000 words, 15ms/1700 words).

The Series One Plus, unlike the Series One and Series One Junior, is compatible with analog I/O. Twenty four of the total I/O can be used for analog applications. In addition to the basic ladder language, the Series One Plus offers advanced data operations, such as direct interfacing with BCD devices. Like the Series One, the Series One Plus can be used in the GEnet Local Area Network. Also, the Series One Plus uses the same programming devices as the Series One.

The Series Three PC is still quite compact, but it has a greater I/O capability, memory, and functionality. The basic Series Three has 4K of CMOS RAM user memory or 4K of EPROM user memory and has 256 I/O points. The system can be expanded to 400 I/O points with the addition of two base units. Of these I/O points, all 400 can be used for discrete and 24 can be used for analog I/O. The Series Three Programmer is built into the basic unit.

Series Three programming is similar to Series One programming. Functions include relay, latch and unlatch, MCR, timer, counter, arithmetical functions, compare functions, and subroutines. The diagnostic functions of the Series Three Programmer help to locate sources of errors in programs.

Scan rates per K of memory vary. Typically, the scan rate for 1K is 12ms; for 2K, 20ms; and for 4K, 35ms.

The Series Three is suitable for use in local area networks (including GEnet). Remote I/O may be distributed in seven locations over a distance of 3000 feet. The RS-422 serial interface is required.

The Series Six family of programmable controllers includes the Model 60, Model 600, Model 6000, and the Series Six Plus. The capacity of each model (in terms of memory and I/O) increases as the model numbers increase. The Series Six Plus, which has the same amount of memory as the Model 6000, has a greater number of I/O points than the Model 6000. All use ladder language and the same programming devices.

The entire Series Six family uses CMOS memory modules. These modules incorporate lithium batteries to retain the contents in memory in the event of a disruption of power. All memory is measured in 16-bit words. The Model 60 has 4K of memory; the Model 600, 8K; the Model 6000 and the Plus, 32K. Typically, the scan rate per K of memory is 1ms.

Total I/O capability for the series is as follows: Model 60, 2000; Model 600, 2000; Model 6000, 4000; Plus, 8000. All of these can be used for discrete I/O, with a smaller number used for analog I/O. The entire Series Six is capable of PID control, and uses the same programming devices as the Series Three. Also, the network capability of the Series Six is the same as the Series Three, with the exception that the Series Six is also compatible in master-slave formats. Communication requires either the RS-232C or RS-422 serial interface.

GENERAL NUMERIC CORP.

General Numeric Corp. of Elk Grove Village, Illinois, offers a variety of German and Japanese PCs. Its smallest PC is the S5-101 U Mini PC. This is a small, low-cost PC suitable for use in the vicinity of power equipment. It allows a total of 64 discrete I/O. It has 1K of memory (0.5K internal RAM with back-up battery and an EEPROM or EPROM plug-in submodule memory). The scan rate per K of memory is 75ms. It can be programmed in ladder language, Boolean, and GRAFCET. Programming is accomplished with the 605 U hand-held programmer or the 675 U CRT programmer. The S5-101 U performs normal mathematical functions.

The S5-115 U PC has 16K of RAM, EPROM, and EEPROM memory. Scan rate per K is 18ms. It has a total of 512 I/O, 64 of which may be used for analog I/O. It can accommodate a maximum of four expansion units. It is capable of PID control.

As with the S5-101 U, the S5-115 U can be programmed with ladder, Boolean, and GRAFCET languages. It uses the same programming devices as the S5-101 U and is capable of performing mathematical and diagnostic functions.

The S5-120 WV PC has 24K of RAM and EPROM memory, with a scan rate per K of 3.8ms. Its I/O capability is much greater than that of the S5-115 U: a total of 1536 I/O points, with 96 usable for analog I/O. It can handle a maximum of four expansion units.

The S5-135 U PC features 32K of RAM and EPROM memory, with a scan rate of 8ms. There are 8192 I/O points, all of which may be used for discrete I/O and 384 of which may be used for analog I/O. The S5-135 U is capable of PID control. It uses the 20mA current loop.

As with the previous PCs, the S5-135 U is programmable in ladder, Boolean, and GRAFCET languages. It uses the same programming devices as the other S5 PCs and is suitable for use in a peer-to-peer local area network. It is capable of performing mathematical and diagnostic functions.

The S5-150 U PC has 48K of RAM and EPROM memories, expandable to 128K. The scan rate per K of memory is 5ms. Like the S5-135 U, it has 8192 total I/O points, with a potential of 384 for use in analog I/O. Communication requires the 20mA current loop.

The S5-150 U features the capability of PID control. The programming languages and devices, mathematical and diagnostic functions, and networking capabilities are similar to the S5-135 U.

The PCs discussed above are German-made. The PC-J, also marketed by General Numeric, is a Japanese-made programmable controller. It features 320K of RAM and EPROM memories with a scan rate of 6ms per program step. It has 1728 I/O points, all of which may be used for discrete I/O, and 108 of which may be used for analog I/O. The PC-J can be programmed with relay ladder language and high-level languages, and has PID control capabilities. Communication requires either the RS-232C or RS-422 serial interface. The PC-J uses 220 volts-ac.

GIDDINGS & LEWIS ELECTRONICS COMPANY

Giddings & Lewis Electronics Company of Fond du Lac, Wisconsin markets two PCs: the PiC 49 and the PiC 409. Giddings & Lewis specialize in motion control, so the PiC 49/409 family is most useful in robotics, material handling, transfer machinery, welding, grinding, and packaging applications.

Both PCs offer 320K of CMOS RAM memory with very fast scan rates per K of memory (0.85ms). The PiC 49 offers 232 I/O points, all of which may be used for discrete I/O and 184 of which may be used for analog I/O. The PiC 409 offers 2032 I/O points, all of which may be used for discrete I/O and 240 of which may be used for analog I/O.

Both PCs can be programmed in ladder language, enhanced ladder language, assembly language, or BASIC. Programming is accomplished either by CRT, cassette tape, or computer. Both PCs are capable of high-speed counting operations, positioning operations, motion control, and PID control.

Both are suitable for use with remote I/O, with a host computer, and in local area networks. Communication requires the RS-422 serial interface, and peer-to-peer communications between PCs can occur over a distance of 4000 feet. Giddings & Lewis offer the NumeriExpress™ Communications Link. Rate of data transmission over this network is selected by the user, up to 76.8 baud. Both PCs perform mathematical and diagnostic functions.

GOULD INCORPORATED

The Industrial Automation Division of Gould Incorporated of Andover, Massachusetts, markets a wide variety of PCs. Two of these are Japanese-made (PC 0085 and PC 0185). The rest are made in America.

The PC 0085 is the smallest Gould PC. It features 1K of CMOS RAM (battery-backed) or EPROM (optional). The scan rate per K of memory is 6ms. The PC 0085 has a nominal 24 I/O points, expandable up to 120 I/O. These are reserved for discrete I/O.

The PC 0085 includes 48 timers and counters and a high speed (5 kHz) counter. It includes diagnostic capabilities. The instruction set for the PC 0085 includes relay, timer, counter, MCR, jump, and compare. It uses a ladder/Boolean language. Programming is accomplished via a hand-held panel that also bolts directly to the PC or by cassette tape or computer. Communication with a host computer or local area network requires the RS-422 serial interface. This low cost PC measures 6 inches high by 5¾ inches wide by 4 inches deep.

The PC 0185 is a mid-sized programmable controller. It features 3.5K of CMOS RAM memory (with battery back-up) or EPROM (optional). The scan rate per K of memory is 2ms. It has a nominal 128 I/O points, expandable up to 512. Using a multidrop configuration, the PC 0185 is compatible with remote I/O. These may be at several different locations and up to 1500 feet from the controller. The PC 0185 is programmed in ladder/Boolean language. Programming is accomplished with the 0110 Programming Panel (a hand-held programmer, which also mounts directly to the PC), cassette tape, or computer. The instruction set includes relay, timer and counter, arithmetical operations, data transfer, AND, OR, XOR, MCR, and jump. It offers a wide range of diagnostic functions. Communication with a host computer or local area network requires the RS-422 serial interface. Neither the PC 0085 or PC 0185 is capable of performing PID control.

The Gould Micro 84 is a slightly larger, low-cost programmable controller. It features 2K (4-bit words) of user memory (EAROM). The scan rate per K of memory is 40ms. It has 112 discrete I/O points, plus 12 analog or BCD I/O points. It is compatible with a large number of Gould I/O modules, which include analog as well as digital devices. It uses the RS-232C serial interface. The Micro 84 is fully compatible with the Gould Modbus local area network. Interfacing with the network requires the J357 communications interface module. With this, the Micro 84 is suitable for use as a slave. Micro 84s can be placed up to a distance of 15,000 feet away from the master. Data transmission in networks of greater than 15,000 feet is possible.

The Micro 84 uses relay ladder language and can be programmed with a hand-held (P370) Program Panel or a P190 CRT Programmer. Additionally, it can be programmed with cassette tape. Once loaded, programs can be transferred to the P371 Portable Program Pack (which uses EAROM). The Micro 84 can perform diagnostic functions.

The Gould 484 PC is designed for mid-sized applications. It is available in five different memory sizes (½K, 1K, 2K, 4K, and 8K). This is CMOS RAM memory. The scan rate per K of memory is 10ms. The 484 features 512 discrete I/O points, and 32 analog I/O points. It uses relay ladder language.

The 484 is compatible with the Gould Modbus. It offers three different instruction sets, PID capabilities, motion control capabilities, and diagnostics.

The Gould 584L PC features memory options that range from 12K to 128K, and scan rates per K of memory between 10 and 60ms. It supports up to 8192 discrete I/O points. It too has PID and motion control capabilities. It is compatible with the Gould Modbus, in which it may serve as master. In addition, the 584L can support remote I/O over a distance of 15,000 feet. Communication requires the RS-232C serial interface. It can be programmed with ladder language, via CRT, cassette tape, or computer. It can perform mathematical operations and diagnostics.

The Gould 884 is a mid-sized PC with 8K of CMOS RAM. The scan rate per K of memory is 25ms. It can support 1024 I/O points. It can be programmed by ladder language, with a set of 35 programming instructions. Communication with Gould's Modbus requires the RS-232C serial interface. The Gould P190 Programmer or the IBM Personal Computer can be used to program the 884. It is capable of PID and positioning control. It can perform mathematical operations and diagnostics.

The Gould 984 PC family is a full-sized PC family. Memory sizes range from 8K to 64K of CMOS RAM (16 or 24 bit words). The scan rate per K of memory is 0.75ms. I/O points range up to 64K (Gould 984D). The 984 family uses relay ladder language, and the instruction set is the same as for the 584L. The 584L family is Modbus-compatible. Communication requires the RS-232C serial interface. The 984 family is capable of performing PID and motion control, mathematical operations, and diagnostics.

GUARDIAN ELECTRIC MANUFACTURING CO.

The Guardian Electric Manufacturing Company of Chicago, Illinois, markets the Hitachi J-16 programmable controller. The J-16 features 2K of EEPROM memory, with a scan rate of 20ms per K. It has a total of 128 discrete I/O points and has remote I/O capabilities. Communication requires either the RS-232C or RS-422 serial interface. The programming language used for the J-16 is relay ladder language. Programming devices include manual, cassette tape, and computer. The J-16 is not suitable for use in a local area network. It does have diagnostic capabilities.

HONEYWELL INC.

The IPC Division of Honeywell (York, Pennsylvania) markets the IPC-620 family of programmable controllers. These are the IPC-620-10, IPC-620-15, IPC-620-20, IPC-620-25, and IPC-620-35. The 10 and 15 models feature 4K of CMOS RAM and EPROM memory, with scan rates of 10ms per K of memory. Both have a total of 512 I/O. The model 15 allows for analog I/O. The model 15 is capable of PID and positioning control. Both are network-compatible (MAP standard). The type of interface used is the RS-232C or RS-422 serial interface. These two PCs are programmed in relay ladder language via CRT, cassette tape, or computer. Both are capable of diagnostics, and the model 15 can perform mathematical functions.

The model 20 has 8K of CMOS RAM, with a scan rate per K of memory of 3.3ms. It too features 512 I/O points (discrete or analog). The model 20 has all the features of the model 15, with the addition of remote I/O capabilities.

Models 25 and 35 feature 32K of CMOS RAM memory, with scan rates per K of memory of 2.5ms. Both have 2048 I/O points (discrete or analog). These two models have all the features of the model 20.

IDEC SYSTEMS AND CONTROLS CORP.

Idec Systems and Controls of Sunnyvale, California offers three PCs: the FA-1, the FA-1J, and the MACH 1. The FA-1 features 1-4K of CMOS RAM, 1-4K of EEPROM, or 4K of EPROM. The scan rate per K of memory is 32ms. The FA-1 has 256 I/O, all of which may be used for discrete I/O and 56 of which may be used for analog I/O. It is programmed with ladder/Boolean language either manually, by cassette tape, or by computer. It uses the RS-232C serial interface. The FA-1 has motion control and diagnostic capabilities.

The FA-1J features 1–4K of CMOS RAM, 1–4K of EEPROM, or 4K of EPROM. The scan rate per K of memory is 32ms. It has 256 I/O, 31 of which may be used for analog I/O. It is capable of PID and motion control, and is programmable with ladder/Boolean language (either manually, by cassette tape, or by computer). The FA-1J can perform mathematical and diagnostic functions.

The MACH 1 features 0.7K of memory, with a scan rate of 40ms per K. It is a smaller PC, offering only 32 discrete I/O. It is programmed with ladder/Boolean language, and can be programmed either manually or with cassette tape. It offers diagnostics, but not mathematical functions.

INDUSTRIAL INDEXING SYSTEMS

Industrial Indexing Systems of Victor, New York offers four programmable controllers: the MM-10, the PSC-100, the MSC-100, and the MSC-800. The MM-10 is the smallest in terms of memory with less than 1K of RAM and ROM memories. The scan rate per K of memory is less than 3ms. It has 34 discrete I/O, and uses a high-level programming language. It is capable of motion control. The MM-10 uses either the RS-232C or the 20mA current loop.

The PSC-100 offers 4K of RAM memory, with a scan rate of K of less than 3ms. It has 24 discrete I/O and can be programmed with a high-level language. The PSC-100 has motion control capabilities. It uses either the RS-232C or 20mA current loop interfaces, as do all the Industrial Indexing Systems PCs.

The MSC-100 features 8K of CMOS RAM memory. It has 40 discrete I/O and can be programmed either with relay ladder or a high-level language. It is capable of PID control.

The MSC-800 has 16K of CMOS RAM, with a scan rate per K of less than 3ms. It has 168 total I/O (72 discrete and 96 analog). It is programmed with a high-level language, and features motion control and PID capabilities.

JOUCOMATIC CONTROLS, INC.

Joucomatic Controls, Inc. of Charlotte, North Carolina distributes the UNITROL M programmable controller. The UNITROL M is a French-made device. It features 1 to 2K of EEPROM memory, with a scan rate per K of memory of 20ms. The UNITROL M has 128 discrete I/O points. The following peripherals are available: a plug-in programmer, analog timer modules, rapid counter modules, and the RS-232C interface module for use with a printer or personal computer.

The UNITROL M is extremely flexible in terms of programming language. It can be programmed with ladder, Boolean, or GRAFCET languages. Functions include AND, OR, NOT, MCR, jump, and latch. The UNITROL M has diagnostic capabilities.

KAYE INSTRUMENTS INC.

Kaye Instruments Inc. of Bedford, Massachusetts distributes the Digistrip 4C and the Digi-Link 4C. These two PCs offer RAM and PROM memory, with a scan rate of eight channels of input per second. Both have a total of 208 I/O, 80 of which are reserved for discrete I/O and 128 of which are reserved for analog I/O. Both can perform PID control, and both use either the RS-232C serial interface or the 20mA current loop.

KLOCKNER-MOELLER CORP.

Klockner-Moeller of Natick, Massachusetts offers the PS family of German-made programmable controllers. The PS3 comes in three versions (PS3-AC, PS3-8, and PS3-DC). The PS3 is a very small, low-cost programmable controller. Its RAM user memory can handle 1K of instructions, with a scan rate per K of less than 5ms. The basic I/O total is 32 points; however, linking as many as four PS3s together results in a maximum I/O of 128. All 128 may be used for discrete I/O, while 20 may be used for analog I/O. System modules include timers, counters, clocks, arithmetical functions, and a high speed counter (10kHz). It uses the RS-485 serial interface.

The PS3 family uses ladder language. Programming is effected either by a hand-held programmer (PRG 3), a light-pen programmer (PRG 300), or with an IBM Personal Computer or an IBM Personal Computer-compatible device. The PS3 is suitable for use in a local area network. When used in a peer-to-peer local area network with other PS3s, the maximum distance between PCs is 600 meters. This is the same maximum distance as when the PS3 is used in a local area network as a slave, with the PS32 as a master.

Klockner-Moeller offers the SUCONET Field Bus Local Area Network. The PS3 performs the four basic arithmetical operations and offers diagnostic capabilities.

The SUCOS PS21, 22, and 24 programmable controllers all feature 4K of EPROM, EEPROM, or RAM memory with scan rates per K of 2ms (1ms for the PS24). The total number of I/O points increase with the model number. Thus the PS21 has a total of 56 discrete I/O; the PS22, 1024 discrete (64 analog) I/O; and the PS24, 2048 discrete (128 analog) I/O. All three use the RS-232C serial interface, and all three are programmed with ladder or Boolean language (manually, or by CRT or computer). All three accept remote I/O.

The SUCOS PS32 offers 23K of memory (up to 16K of instructions in RAM and 16K in EPROM). Scan rate per K of memory is 2.5 ms. The PS32 has 2048 total I/O. All of these may be used for discrete, and 128 may be used for analog I/O. It can be programmed in ladder and Boolean languages (manually or with a computer). The RS-232C, RS-422, and RS-485 serial interfaces are used by the PS32. It is capable of PID and positioning control. The PS32 is suitable for use as a master in the SUCONET Field Bus local area network, or in peer-to-peer formats. It performs mathematical and diagnostic operations.

LEHIGH FLUID POWER

The Miyashita TPC-20, marketed by Lehigh Fluid Power (Lambertville, New Jersey), was mentioned in Chapter 10. The TPC-20 features 470 words of CMOS RAM memory with a scan rate of 20ms per K of memory. It has 20 discrete I/O points, and is capable of supporting remote I/O. It is suitable for use in a MAP local area network. The TPC-20 has high-speed counting and PID capabilities. It is programmed with a mixture of ladder and high-level languages.

MTS SYSTEMS CORPORATION

MTS of Minneapolis, Minnesota, offers the Motion Plus 473 and Motion Plus TDC series of programmable controllers. The 473 has 32K of RAM and EPROM memories. It has 256 I/O, all of which may be used for discrete I/O, and 24 of which may be used for analog I/O. The 473 can be programmed with high-level language and offers motion control, high-speed counting, positioning, and PID control capabilities. Also, it can perform mathematical and diagnostic operations. It can be programmed with a computer and uses the RS-232C serial interface.

The TDC PC offers 700 steps of EPROM and RAM memory, with a scan rate of 2ms per K. It is smaller than the 473 PC: it has a total of 22 I/O points. It is programmable with a high-level language via a computer, and offers positioning and PID control capabilities as well as diagnostics. It uses the RS-232C serial interface. Like the 473, the TDC is suitable for use in a peer-to-peer local area network.

MAXITRON CORP.

Maxitron Corp. of Corte Madera, California offers the PLC and DPC lines of programmable controllers. The PLC 47 JR has 16K of RAM and EPROM memories. It has a total of 80 I/O (discrete and analog). The PLC 47 JR can be programmed in ladder, Boolean, and GRAFCET languages manually or with a computer. It is suitable for use in a local area network, and can perform mathematical and diagnostic functions.

The PLC 47 has 16K of RAM and EPROM memories and a total of 256 discrete and analog I/O points. It is programmed in the same languages and with the same devices as the PC 47 JR. It is suitable for use in a local area network and can perform mathematical and diagnostic functions.

The PLC 67-30 has 32K of RAM and EPROM memories and a total of 512 I/O points (analog and discrete). It has high-speed counting, positioning, and PID control capabilities. It can be programmed in the same language as the PLC 47, via a CRT or computer. The PLC 67-30 has remote I/O and network capabilities, and performs mathematical and diagnostic functions.

The PLC 87-10 has 128K of RAM and EPROM memories with 992 I/O points. It is capable of high-speed counting, positioning, and PID control. It is programmed with the same languages as the PLC 67-30, via a CRT or computer. It offers remote I/O as well as networking capabilities. It can perform mathematical and diagnostic functions.

The PLC 87-20 is very similar to the PLC 87-10, except that it has 2016 I/O points.

The DPC 67-30 has 64K of RAM and EPROM memories, with a scan rate per K of memory of 0.45ms. It offers 512 I/O points, all of which can be used for discrete I/O and 32 of which can be used for analog I/O. It can perform high-speed counting, positioning, and PID control. It is programmed with ladder, Boolean, and GRAFCET languages via a computer. Networking capabilities include remote I/O and MAP local area networks. It can perform mathematical and diagnostic functions. It uses the RS-232C, RS-422, and RS-449 serial interfaces.

The DPC 87-10 has 128K of RAM and EPROM memories, with a scan rate per K of 0.45ms. Total I/O capability is 992, all of which may be used for discrete and 60 of which may be used for analog I/O. It can be programmed with the same languages as the 67-30, via a computer. Its network capabilities are the same as the 67-30. It has high-speed counting, positioning, and PID control capabilities. It uses the same interfaces as the DPC 67-30.

The DPC 87-20 is similar to the DPC 87-10, except that it has a greater number of I/O points (2016, 124 of which may be used for analog I/O).

MCGILL MANUFACTURING CO.

McGill Manufacturing Co. of Valparaiso, Indiana offers three programmable controllers: the 1701-1000, the 1701-2000, and the 1701-7000. The 1000 model has 4K of RAM and EPROM memories with a scan rate per K of 8.3ms. it offers 512 I/O, all of which may be used for discrete I/O and 256 of which may be used for analog I/O. It has high speed counting and position control capabilities. It is programmable in ladder and Boolean languages, via a computer. it is suitable for use in a peer-to-peer format local area network. It offers diagnostic functions and uses the RS-232C serial interface.

The 1701-2000 has 4K of RAM and EPROM memories with a scan rate per K of memory of 8.3ms. Like the 1701-1000, the model 2000 offers 512 I/O. Its other capabilities are similar to those of the 1701-1000.

The 1701-7000 offers 8K of EPROM memory and 512 I/O points. In other respects it is similar to the 1701-2000.

MINARIK ELECTRIC CO.

Minarik Electric Co. of Los Angeles, California offers the Micromaster LS programmable controller. It has 2878 words of RAM memory with a scan rate of 7ms per K. It features 176 I/O: 120 discrete and 56 analog. It can be programmed with ladder or high-level languages.

MITSUBISHI ELECTRIC

The Industrial Automation Division of Mitsubishi Electric (Mt. Prospect, Illinois) offers the K, F, and A families of PCs. The K family includes the K0J, K2N, and K3N PCs. They offer 2K, 4K, and 16K, respectively, of RAM and EPROM memories. Scan rates per K of memory are 5.6, 3.8, and 1ms, respectively. Total I/O capabilities are: K0J—184 total (160 discrete, 46 analog); K2N—512 total (512 discrete, 128 analog); and K3N—2048 total (2048 discrete, 512 analog). All three offer high-speed counting, positioning, and PID control capabilities. All three are programmed with ladder and Boolean languages, either manually, by CRT, by cassette tape, or by computer. The K2N and K3N offer remote I/O capabilities. The K3N is suitable for use in local area networks. All three offer mathematical and diagnostic functions.

The F series includes the F-12, F-20, F2-40, and F2-60. All four have CMOS RAM memory. In addition, the F-20, F2-40, and F2-60 have UV-PROM memory. Memory sizes are as follows: F-12, 320 steps; F-20, 320 steps; F2-40, 422 steps; F2-60, 422 steps. Scan rates per K are F-12, 100ms; F-20, 45ms; F2-40, 7ms; F2-60, 7ms. All four use ladder and Boolean languages. All four can be programmed manually, by CRT, by cassette tape, or by computer. Additionally, all four offer diagnostic functions.

Total I/O capabilities for the F series are as follows: F-12, 32; F-20, 40; F2-40, 80; and F2-60, 120. All of these are reserved for discrete I/O. The F2-40 and F2-60 are capable of high speed counting operations, while all four are capable of positioning operations. The F2-40 and F2-60 are suitable for use in a peer-to-peer local area network and can perform mathematical functions. All four use the RS-422 serial interface.

The A series includes the A1, A2, and A3. All of these use RAM, EPROM, and EEPROM memories. The A1 has 14K, while the A2 and A3 have 144K of memory. I/O totals are as follows: A1, 256; A2, 512; A3, 2048. All three have high speed counting, positioning, and PID control capabilities. All three can be programmed in ladder/Boolean and high-level languages. All three can be programmed either manually, by CRT, by cassette tape, or by computer. They all offer remote I/O capabilities, as well as compatibility with MAP local area networks. Also, they all perform mathematical and diagnostic functions.

MODULAR COMPUTER SYSTEMS, INC.

Modular Computer Systems, Inc. of Fort Lauderdale, Florida offers the APC/5 programmable controller. It features 2M of RAM memory with 2048 I/O. Of these, all may be used for discrete I/O and 1280 may be used for analog I/O. It can be programmed in ladder and high-level languages and has PID and motion control capabilities. It uses the RS-232C serial interface.

NAVCOM, INC.

Navcom, Inc. of Huron, Ohio offers five programmable controllers: the F10A, the F20A, the F26A, the F36B, and the F40A. All of these feature 32K of EPROM and RAM memories, with scan rates per K of memory of 40ms. All use the RS-232C, RS-422, and RS-485 interfaces.

The F10A has a total of 10 I/O (all discrete). It can be programmed with relay ladder language and offers motion control capabilities.

The F20A has 20 discrete I/O. It is also programmed with relay ladder language and has motion control capabilities. The F26A is similar to the F20A, except that it has 26 I/O, all of which may be used for discrete and 16 of which may be used for analog I/O. The F36B is similar to the F26A except that its total I/O capability is 36 (36 discrete I/O, 16 analog I/O). The F40A is similar to the F36B, except that it offers 40 discrete I/O.

OMRON ELECTRONICS, INC.

Omron Electronics, Inc. of Schaumburg, Illinois offers a full line of programmable controllers under the SYSMAC name. These include the SYSMAC S6, the SYSMAC C20, the SYSMAC C120, and the SYSMAC C500.

The SYSMAC S6 is a small, inexpensive programmable controller that features 512K of RAM and EPROM memory with a scan rate per K of 10ms. It has 64 discrete I/O. It can be programmed in ladder/Boolean language. Programming is accomplished with the use of the SYSMAC PRO 16 Programming Console.

The SYSMAC C series of programmable controllers includes the C20, C120, and C500 PCs. These have 1194, 2.6K, and 8K of RAM and EPROM memories, respectively. Scan rates per K are 10ms (C20 and C120) and 5ms (C500). The SYSMAC C20 uses a fiberoptic interface, and the C120 and C500 use either the RS-232C or RS-422 serial interface. The C20 has a total of 140 discrete I/O points and uses relay ladder/Boolean language. It can be programmed manually, by CRT, by cassette tape, or by computer. The C20, C120, and C500 offer remote I/O capabilities. They can be used in peer-to-peer local area networks. All three can perform diagnostic and mathematical functions.

The C120 and C500 offer 256 and 512 I/O points, respectively. All of these may be used for discrete I/O. The SYSMAC C120 has 32 analog I/O, while the SYSMAC C500 has 64 analog I/O. Both use the same programming language as the SYSMAC C20. The C500 is capable of PID and motion control. The C120 and C500 are suitable for use in local area networks (master-slave format). Both the C120 and C500 are compatible with the SYSBUS local area network. Both are suitable for use with remote I/O.

PHOENIX DIGITAL CORP.

Phoenix Digital Corp. of Phoenix, Arizona offers the DPAC 80 programmable controller. This PC has 180K of CMOS RAM and UV-PROM memories, with a variable scan rate per K of memory. It offers 832 total I/O, 320 of which are for discrete and 512 of which are for analog I/O. It uses the relay ladder language and high-level language. The DPAC 80 offers PID control capabilities. It uses the RS-232C, RS-422, and fiberoptic interfaces. The DPAC 80 can be programmed with a CRT or a computer.

POLYMONT CORP.

Polymont Corp. of Dearborn, Michigan offers the SMC 25, SMC 50, SMC 80, and PB 600 programmable controllers. These are all French-made PCs. The SMC 25 has 4K of EEPROM memory with 304 discrete I/O. It can be programmed with ladder or high-level languages. The SMC 25 uses the 20mA current loop interface.

The SMC 50 has 8K of RAM and EPROM memories, with 228 discrete I/O. Programming languages include ladder and high-level. It has motion control capabilities and uses either the 20mA current loop, RS-232C, or RS-422 interfaces.

The SMC 80 offers 2K of EPROM memory with a scan rate per K of less than 3.5ms. It has 320 discrete I/O. It can be programmed in ladder or high-level language. It uses the same interfaces as the SMC 50.

The PB 600 offers 32K of RAM and EEPROM memories. A total of 2048 I/O points are available, all of which may be used for discrete and 256 of which may be used for analog I/O. It can be programmed in ladder and high-level language. Motion control capabilities are included. It uses the same interfaces as the SMC 50 and SMC 80.

RELIANCE ELECTRIC COMPANY

The Control Systems Division of Reliance Electric Company of Stone Mountain, Georgia, offers the DCS 5000 and the AutoMate series of PCs. The DCS 5000 features 80K of RAM memory, with 1152 I/O points. All these I/O points may be used for discrete I/O, while 128 may be used for analog I/O. It can be programmed with ladder/Boolean and high-level languages. It is capable of positioning and PID control. It uses the RS-232C serial interface. (This interface is also used with the AutoMate series.) The DCS 5000 can be programmed with a CRT or computer. It is suitable for use in a MAP protocol local area network (peer-to-peer or master-slave formats). It can perform mathematical and diagnostic functions.

The AutoMate series includes the AutoMate 15, 20, 30, and 40. All of these have RAM memory. In addition, the AutoMate 15 features NOVRAM memory. Memory size and scan rate per K of memory are as follows: 15—1K at 4ms; 20—2K at 4ms; 30—8K at 2 ms; 40—104K at 0.8ms. Total I/O points vary throughout the series. The AutoMate 15 has the fewest I/O points (64, all discrete), while the AutoMate 40 has the most I/O points (8704, 8192 discrete and 512 analog). The AutoMate 20 has 256 discrete I/O, while the AutoMate 30 has 640 total I/O (512 discrete and 128 analog). The AutoMate 30 and AutoMate 40 have motion and PID control, and high-speed counting capabilities. All use relay ladder language. The series can be programmed by CRT or computer. All but the AutoMate 15 offer remote I/O capabilities, and all are suitable for use in local area networks. All but the AutoMate 15 can perform mathematical operations, and all offer diagnostics.

SIEMENS ENERGY & AUTOMATION, INC.

Siemens Energy & Automation, Inc. of Peabody, Massachusetts offers the S5 series of programmable controllers. These include the S5-100 U, S5-101 U, S5-101 R, S5-105 R, S5-115 L, S5-115 U, S5-135 U-S, S5-135 U-R, and S5-150 U.

The S5-100 U has 4K of RAM, EPROM, and EEPROM memories with a scan rate of 7ms per K. It has a maximum of 272 I/O points, 256 discrete and 16 analog. It uses relay ladder programming language and has PID and motion control capabilities.

The S5-101 U has 1K of RAM, EPROM, and EEPROM memories with a scan rate of 30ms per K. It features a total of 64 discrete I/O. It is programmed with ladder/Boolean language, either manually or with a CRT. It is suitable for use in a MAP-compatible local area network. It performs mathematical and diagnostic functions.

The S5-101 R has 384 words of RAM, EPROM, and EEPROM memories with a scan rate of 2.5ms per K. It has 32 discrete I/O points and is programmed with ladder language. It is programmed manually and offers mathematical and diagnostic functions.

The S5-105 R has 1K of RAM, EPROM, and EEPROM memories with a scan rate of 5ms per K. It has a total of 128 discrete I/O. It can be programmed with ladder language, either manually or with a CRT. It offers mathematical and diagnostic capabilities.

The S5-115 L has 6K of RAM and EEPROM memories with a scan rate of 6ms. It offers a total of 976 I/O points (976 discrete or 128 analog). It is programmed with relay ladder language. It uses the RS-232C serial interface. The S5-115 L is suitable for use in MAP-compatible local area network and can be used with remote I/O. It performs mathematical and diagnostic functions.

The S5-115 U features 20K of RAM, EPROM, and EEPROM memories with a scan rate of 5ms per K. It has 2048 I/O points (2048 discrete or 128 analog). It can be programmed with ladder/Boolean language and is capable of positioning and PID control. It uses the RS-232C interface. Its networking capabilities are the same as the S5-115 L. It performs mathematical and diagnostic functions.

The S5-135 U-S features 128K of RAM, EPROM, and EEPROM memories with a scan rate of 2ms per K. It offers 8192 I/O points (8192 discrete and 384 analog). It has high-speed counting, positioning, and PID control capabilities. It is programmed with ladder/Boolean language, either manually or with a CRT. It is suitable for use with remote I/O and MAP networks. It can perform mathematical and diagnostic functions.

The S5-135 U-R features 8K of RAM and EPROM memories, and 8192 I/O points (8192 discrete, 384 analog). Its programming language, PID and positioning control capabilities, and network compatibility are the same as for the S5-135 U-S PC.

The S5-150 U PC features 64K of RAM and EPROM memories, with a scan rate of 1.5ms per K. It offers 38K total I/O points (38K discrete, 384 analog). It can be programmed with ladder/Boolean and GRAFCET languages. It can be programmed either manually or with a CRT. It is suitable for use in MAP local area networks and has remote I/O capabilities. It uses the RS-232C serial interface.

SOLID CONTROLS, INC.

Solid Controls, Inc. of Minneapolis, Minnesota distributes the System 10, the EPIC 1, and the EPIC 8B programmable controllers. The System 10 features 16K of EPROM memory at a scan rate of 2.5ms per K. It has 136 total I/O points (128 discrete, 8 analog). It is programmed with Boolean language. It offers high-speed counting and positioning control capabilities. It can perform diagnostic functions.

The EPIC 1 features 16K of EPROM memory at a scan rate of 2.5ms per K. There are a total of 520 I/O points (512 discrete, 8 analog). The EPIC 1 has high-speed counting and positioning control capabilities and is programmed with Boolean by a computer. It performs diagnostic functions.

The EPIC 8B features 392K of EPROM and RAM memories, at a scan rate per K of 1.5ms. It has a total of 584 I/O points (384 discrete, 200 analog). It has high-speed counting, positioning, and PID control capabilities. It is programmed with Boolean and a high-level language by a computer. It uses the RS-232C and RS-422 interfaces. It is suitable for use with remote I/O and in local area networks. It performs mathematical and diagnostic functions.

SQUARE D CO.

The Automation Products Division of Square D Co. (Milwaukee, Wisconsin) offers the SY/MAX series of programmable controllers. These include the SY/MAX Model 100, Model 300, Model 500, and Model 700.

The SY/MAX Model 100 is a British-made PC that features 420 words (16 bit) of memory. There are two different types of memory available: RAM or UV-PROM. The scan rate per K of memory is 10ms. The Model 100 also offers two choices in I/O points (20 or 40). These are all discrete I/O. It is programmed with ladder language, the instruction set of which includes relays, latch and unlatch, data comparison, timers, counters, and MCR. It can be programmed manually, by CRT, cassette tape, or computer. It is suitable for use in a local area network (the SY/NET local area network) with the use of the Class 8030-type Network Interface Module. SY/NET follows the MAP protocol. The RS-422 serial interface allows communication with peripherals to occur. It offers diagnostic functions.

The Model 300 features 2K of RAM or UV-PROM memories with a scan rate per K of 30ms. A mixture of RAM and UV-PROM is also available as an option. This U.S.-made PC has 256 I/O points (256 discrete, 112 analog). It is programmed with ladder language, either manually or by CRT, cassette tape, or computer. It uses the RS-422 serial interface. The Model 300 has high speed counting, positioning, PID control, and remote I/O capabilities. It can be used with the SY/NET local area network. It offers mathematical and diagnostic capabilities.

The Model 500 has 8K of RAM or RAM/UV-PROM memories with a scan rate of 2.6ms per K. Total I/O capabilities includes 2000 points, all discrete or 1785 analog.

It features all the functions of the Model 300 (control, networking, etc.) with additional advanced capabilities. These include expanded mathematical functions (square root, random number, absolute value, etc.), scan control (GOTO, subroutines, etc.), matrix operations, immediate communications update, Boolean functions (AND, OR, XOR), etc. It has PID control and positioning capabilities and uses the SY/NET local area network. Communication between the Model 500 and remote I/O modules can occur over a distance of 15,000 feet. The Model 500 is programmed the same way as the Model 300 (manually, CRT, cassette tape, or computer). Communication with peripherals requires the RS-422 serial interface.

The Model 700 features 64K of RAM memory. It offers a Memory Control Unit (MCU) which includes nonvolatile back-up memory (bubble memory). Scan rate per K of memory is 1.3ms. The Model 700 has 14K total I/O points (all discrete or 1785 analog). The ladder language instruction set includes relays, latch and unlatch, MCR, timers, counters, data comparison, binary to BCD, BCD to binary, mathematical operations (including square root, absolute value, sine and cosine, base 10 logarithms, natural logarithms, power functions, and advanced trigonometric operations), PID control, and Boolean operations. It is also capable of motion control (positioning). It can be programmed manually, by CRT, cassette tape, or computer. Like the other SY/MAX models, the Model 700 uses SY/NET, and the RS-422 serial interface.

STRUTHERS-DUNN INC.

The Systems Division of Struthers-Dunn Inc. of Bettendorf, Iowa, distributes the Director 4001, the Director 4002, and the Director One. The Director 4001 has 6K of RAM and EEPROM memories with a total of 384 I/O points (discrete or analog). It offers high-speed counting, remote I/O, positioning, and PID control capabilities. Ladder programming language is used by the 4001. It can be programmed by CRT, cassette tape, or computer. It is suitable for use in a local area network and performs mathematical and diagnostic functions.

The Director 4002 is similar in functionality to the 4001, minus the local area network capabilities. The 4002 offers 6K of RAM and EEPROM memories with a total of 64 I/O points (discrete or analog).

The Director One offers 1970 steps of RAM or EEPROM memory. Like the 4001 and 4002, it has a scan rate per K of 20ms. A total of 128 I/O points are available (discrete or analog). Programming is with ladder language and Boolean, either manually, by CRT, cassette tape, or computer. It has remote I/O and diagnostic capabilities.

All these Director PCs use either the RS-232C or RS-422 serial interfaces.

TELEMECANIQUE INC.

Telemecanique Inc. of Westminster, Maryland markets a variety of French- and Japanese-made programmable controllers. These include the MPC-007 (Japanese—the remaining PCs are French-made), the TSX-27, the TSX-47-JR, the TSX-47, the TSX-67-30, the TSX-87-10, and the TSX-87-20.

The MPC-007 features 4K of RAM and EPROM memories with a scan rate of 32ms per K. It has 256 I/O points total (256 discrete or 32 analog). It uses ladder/Boolean programming language and can be programmed either manually or by computer. It is suitable for use in a peer-to-peer local area network and performs both mathematical and diagnostic functions.

The TSX-27 features 32K of CMOS RAM and EPROM memories with a scan rate per K of 2ms. It has 80 discrete I/O points. It uses ladder, Boolean, and GRAFCET programming languages. The 20mA loop is used in communications. It can be programmed either manually or by cassette tape and offers mathematical and diagnostic functions.

The TSX-47-JR offers 32K of CMOS RAM and EPROM memories with a scan rate per K of 2ms. It has a total of 80 discrete I/O points and 224 analog I/O points. It is capable of high-speed counting and positioning operations. It uses ladder, Boolean, and GRAFCET programming languages. It can be programmed either manually or by cassette tape. The TSX-47-JR, like the MPC-007, is suitable for use in peer-to-peer local area networks and performs mathematical and diagnostic functions. It uses the RS-232C or RS-422 serial interface.

The TSX-47 programmable controller also has 32K of CMOS RAM and EPROM memories, with the same scan rate as the TSX-47-JR. It offers 256 I/O points (256 discrete or 44 analog). In other respects it is similar to the TSX-47-JR.

The TSX-67-30 has 32K of CMOS RAM and EPROM memories with a scan rate per K of 0.5ms. It has 512 I/O points (512 discrete, 64 analog) and is programmable with ladder and GRAFCET languages. It offers high-speed counting and PID control capabilities. It is suitable for use in peer-to-peer format local area networks and with remote I/O. It too performs mathematical and diagnostic functions.

The TSX-87-10 has 64K of CMOS RAM and EPROM memories with a scan rate per K of 0.5ms. It offers 1024 total I/O points (128 analog). It is otherwise similar to the TSX-67-30.

The TSX-87-20 is similar to the TSX-87-10, with the exception that the TSX-87-20 has 128K of memory and 2048 total I/O points (256 analog).

TENOR CO. INC.

Tenor Co. Inc. of New Berlin, Wisconsin produces the 100 and the PSC-763 programmable controllers. The 100 is made in the U.K. and has 32K of CMOS RAM and EPROM memories with a scan rate a 10ms. It features 252 I/O points (all discrete or 15 analog). It uses relay ladder and high-level programming language. It is capable of motion and PID control and uses the RS-232C, RS-422, or RS-485 serial interfaces.

The PSC-763 offers 128 steps of EPROM memory, and 96 discrete I/O points. It is programmed with a high-level language.

TEXAS INSTRUMENTS

The Industrial Controls Department of Texas Instruments of Johnson City, Tennessee, offers a wide variety of programmable controllers. These include the 510, 520, 5TI, TI-100, 530C, PM-550C, and 560/565.

The 510 offers 256 words of CMOS RAM and EPROM memories. It has 40 discrete I/O and is programmed in relay ladder/Boolean language. It can be programmed manually, by CRT, or by cassette tape. it offers diagnostics and uses the RS-236 interface. Like all TI programmable controllers, it is suitable for use with the TIWAY ONE local area network.

The 520 offers 2K of CMOS RAM and EPROM memories. It supports 128 I/O points (discrete or analog) and is programmable with ladder and high-level languages. It has motion and PID control capabilities. It uses the RS-232C or RS-422 serial interfaces. It is suitable for use in a local area network and performs mathematical and diagnostic functions.

The 5TI has 4K of RAM and EPROM memories with a scan rate per K of 8.3ms. It offers a total of 512 discrete I/O points and is programmable with ladder language. It uses the RS-232C serial interface. The 5TI can be programmed manually or with a CRT. It is suitable for use in a local area network and offers diagnostic capabilities.

The TI-100 has 1K of RAM and EPROM memories with a scan rate of 5ms per K, and a total of 128 discrete I/O points. It can be programmed with ladder language either manually or by cassette tape. It uses the RS-232C serial interface.

The 530C series features between 7 and 19K and RAM and EPROM memories with a scan rate per K of 4ms. The series has 1023 I/O points (discrete or analog). Programming is with ladder language. This series features motion control, high-speed counting, and PID control capabilities. It uses the RS-232C, RS-422, or RS-423 interfaces. It is suitable for use in the TIWAY ONE local area network. It can perform mathematical and diagnostic functions.

The PM-550C series features 4 to 7K of EPROM, EAROM, and RAM memories. Total I/O capability is 640 points (512 discrete, 128 analog). The programming language is relay ladder, and programming can be done manually or with a CRT. It has high-speed counting, motion control (positioning), and PID control capabilities. It uses the RS-232C or RS-422 serial interfaces. It is suitable for use with the TIWAY ONE local area network, and offers mathematical and diagnostic functions.

The 560/565 programmable controller offers 256K of CMOS RAM memory with a scan rate of 2.2ms per K. It has a total of 8192 I/O points (discrete or analog). The programming language is ladder. The 560/565 features positioning, high-speed counting, and PID control capabilities. Programming is accomplished either with a CRT or a computer. It uses the RS-232C, RS-422, or RS-423 interfaces. It can be used in the TIWAY ONE local area network and performs mathematical and diagnostic functions.

TEXAS INSTRUMENTS

The Texas Instruments group of Dublin, California, which was formerly ETI Micro, offers the 8630, 8640, and 8641 series of programmable controllers. The 8630 has 64K of RAM, EPROM, and EEPROM memories, with 52 I/O points (32 discrete, 20 analog). It can be programmed either with ladder or high-level language. It has PID and motion control capabilities, and uses the RS-422 serial interface (as do the other two PCs).

The 8640 features 110K of RAM, EPROM, and EEPROM memories. There are a total of 44 I/O points (34 discrete, 10 analog). In other aspects, it is similar to the 8630.

The 8641 is similar to the 8640 except that it has 256 I/O points (256 discrete, 128 analog).

THESAURUS, INC.

Thesaurus, Inc. of Huntsville, Alabama markets four programmable controllers: the CBPC-1, the CBPC-2, the CBPC-3, and the CBPC-4. Model 1 features 500K of RAM memory with a scan rate of 0.5ms per K. A total of 256 I/O points are available (discrete or analog). It uses either ladder or high-level languages and has motion and PID control capabilities. It uses the RS-232C, RS-422, or RS-488 interfaces.

The Model 2 features 1M of RAM memory with a scan rate per K of 0.2ms. It offers 512 I/O points (discrete or analog). In other respects it is similar to the Model 1.

The Model 3 features 2M of RAM memory with a scan rate of 0.1ms per K. It offers 1024 I/O points (discrete or analog). In other respects it is similar to the Model 1.

The Model 4 features 16M of RAM memory with a scan rate per K of memory of 0.01ms. It has 2048 I/O points (discrete or analog). In other respects it is similar to the Model 1.

TOSHIBA/HOUSTON

Toshiba of Houston, Texas, offers the EX series of programmable controllers. The EX-20/40/40H series is composed of high performance PCs that use ladder language. Standard features include a built-in PROM writer, a 120 volt ac power supply, and I/O status LEDs. The series is designed for small-scale operations, including I/O from 40 to 120 points. The EX-20 features 1K of CMOS RAM and EEPROM memories with a scan rate per K of 60ms. It has 40 total I/O (discrete or analog). It uses the RS-422 interface. The EX-40 features 1K of CMOS RAM and EEPROM memories, at the same scan rate as the EX-20, and a total of 80 I/O (discrete or analog). In other respects it is similar to the EX-20. The ES-40H features 2K of memory, a faster scan rate (3ms), and 120 total I/O.

The EX-200B features 4K of CMOS RAM and EEPROM memories with a total of 752 I/O points. It is capable of communication with personal computers through its built-in RS-422 serial interface port. The EX-200B can be programmed with an IBM personal computer. It includes 60 special functions (compare, trigonometry, mathematical functions, timers, counters, etc.). The RS-422 allows the EX-200B to be used in a local area network (multidrop common bus configuration). It also offers diagnostic functions and is compatible with the EX-250 and EX-500.

The EX-250 and EX-500 programmable controllers are similar to each other in many respects. Both feature CMOS RAM and EPROM memories, both use the RS-422 serial interface, both are programmable with ladder language, and both feature motion and PID control capabilities. Both are suitable for use in local area networks and with remote I/O. Both offer mathematical and diagnostic functions. The differences between the two are as follows. The EX-250 has 4K of memory, a scan rate per K of 3.5ms, and 512 I/O points (512 discrete or 32 analog). The EX-500 has 8K of memory, a scan rate per K of 2.9ms, and 1024 I/O points (1024 discrete or 64 analog).

TRICONIX CORPORATION

Triconix of Irvine, California, offers the TRICON programmable controller. It has 144K of RAM and PROM memories, with scan rates per K of 20 to 70ms. It features 1024 I/O points (discrete or analog). It is programmable with ladder/Boolean, and high-level languages. It is capable of PID control and is programmed by a CRT or computer. The TRICON is suitable for use with remote I/O and in local area networks. It uses the RS-232C or RS-422 interfaces. It can perform mathematical and diagnostic functions.

VEEDER-ROOT CO.

The Digital Products Division of Veeder-Root of Hartford, Connecticut, markets the V-12 Standard and the V-12 Expanded (EXP) programmable controllers. The V-12 Standard features 944 words of CMOS RAM and EPROM memories with a scan rate of 40ms. This Japanese-made PC has 120 I/O points (120 discrete, 15 analog). It is programmed with ladder/Boolean language either manually or by cassette tape. It performs diagnostics.

The V-12 EXP also uses CMOS RAM and EPROM memories, but with 832 word of memory at a slower scan rate (45ms). It features 80 I/O points (80 discrete, 8 analog). It is programmed with ladder/Boolean language. The EXP offers high-speed counting and positioning capabilities. In other respects, it is similar to the V-12 Standard.

WESTINGHOUSE ELECTRIC CO.

The Programmable Controls Department of Westinghouse of Pittsburgh, Pennsylvania, markets several lines of programmable controllers. These are characterized by size as PICO, MICRO, SMALL, MEDIUM, MEDIUM/LARGE, and HIPPO.

The PICO PCs include the Numa-Logic PC-100 and PC-110. Both are designed to replace relays in small applications. The PC-100 can replace more than 75 relays, counters, and timers. It features 320 eight-bit words of user memory (RAM and EPROM). The scan rate per K of memory is 8ms. It has a total of 30 discrete I/O points. It uses ladder-based Boolean mnemonic programming language and can be programmed either manually with the NLPL-180 hand-held program loader or with the NLPL-1581 (IBM Personal Computer). The instructions set includes AND, OR, NOT, MCR, relays, latch and unlatch, timer, and counter. The RAM memory is supported by an internal capacitor or battery. Neither the PC-100 nor the PC-110 are suitable for use with special or remote I/Os.

The PC-110 features a memory size of 1K (RAM and EPROM). The scan rate per K is 20ms. It has a total of 112 discrete I/O. The programming language is the same as that used with the PC-100. The PC-110 can be programmed with the same devices as the PC-100. It is otherwise similar to the PC-100.

The Westinghouse MICRO PC is the Numa-Logic PC-1100. It is very small in size, but is expandable with various I/O modules. It features up to 3.5K (16 bit words) of RAM and EEPROM memories. The scan rate per K of memory is 7ms. It has 128 total I/O points (discrete, or 16 analog). Remote I/O communications can occur at a rate of 32 Kbaud over a maximum distance of 4000 feet. The programming language is relay ladder, which includes standard functions (relay, latched and unlatched, timer, counter, MCR, arithmetical, comparison, and BCD/binary conversions), and advanced functions (AND, OR, XOR, ASCII, and square root). In addition to the standard and advanced functions, the PC-1100 performs PID control. It can be programmed by the NLPL-1581 (IBM Personal Computer), the NLPL-789, and the NLPL-780 programmers. The PC-1100 is ideally suited for use with the Westnet II local area network (data highway). This local area network is built-in; i.e., it is standard with the advanced instruction set. It allows communications up to 10,000 feet via coaxial cable at 1 Mbaud, among 50 PCs. It has a built-in RS-232C serial interface port.

The Numa-Logic PC-900 is Westinghouse's SMALL PC, and the Numa-Logic PC-700 is Westinghouse's MEDIUM PC. These two share the same peripherals, communication formats and protocols. The PC-900 features a maximum 2.5K of RAM memory and a scan rate per K of 20ms. It offers a total of 256 I/O points (256 discrete, 32 analog). Like the PC-1100, the PC-900 supports special I/O modules (such as high-speed counter and ASCII). Communication with remote I/O can occur over a distance of 10,000 feet. The PC-900 has all the standard and advanced programming instructions that are used with the PC-1100. It is programmed by the same devices as the PC-1100 and uses the RS-232C serial interface.

The Numa-Logic PC-700 features a larger memory than the PC-900 (up to 8K of RAM). Its scan rate per K of memory is 8ms. It has a total of 512 discrete I/O points and 64 analog I/O points. The instruction set includes relay, latch and unlatch, timer, counter, MCR, skip, arithmetical functions, comparison, BCD/binary conversion, AND, OR, NOT, ASCII, square root, and PID control.

Westinghouse markets three MEDIUM/LARGE PCs. These are the Numa-Logic MAC series PCs, the MAC-4000, MAC-4010, and MAC-4500. These offer from 10 to 64K of RAM memory, with 1.2ms scan rates per K. The I/O capability for the series is a total of 8192 points (discrete and analog). All are programmed with ladder language and use the RS-232C serial interface. Like all the previously mentioned Westinghouse PCs, the MAC series PCs can be used on the Westnet II data highway. All three offer advanced functions, and the MAC-4500 offers the PID algorithm. Communication with remote I/O over a distance of 10,000 feet can occur at the rate of 865 Kbaud.

The Westinghouse HIPPO PCs are the Numa-Logic HPPC-1500 and HPPC-1700 programmable controllers. These are full-size, extremely powerful PCs. Both offer up to 224K of RAM memory with a scan rate per K of 1ms. Both have a total I/O capability of 8192 points in any mix of discrete or analog. All special functions and special I/O modules mentioned previously are included in the HPPC series capabilities (e.g., high speed counting, PID control, motion control, positioning, etc.). Both can be used with the

Westnet II data highway. Both are programmed by computer (IBM Personal Computer). The programming language is relay ladder.

SUMMARY

The preceding data on programmable controllers should provide you with a broad overview of the diversity of PCs available today. The data presented in this chapter should be confirmed with the manufacturer, since changes in the programmable controller industry occur quite rapidly.

Appendix A

Domestic Suppliers of Programmable Controllers

ASEA Industrial Systems
P.O. Box 372
Milwaukee, WI 53201
414/785-3200

Adatek, Inc.
P.O. Box 1339
Sandpoint, ID 83864
208/263-1471

Allen-Bradley Co.
1201 S. Second St.
Milwaukee, WI 53204
414/382-2000

Anderson Cornelius Co.
6750-T Shady Oak Rd.
Eden Prairie, MN 55344
612/944-3220

Applied Systems Corp.
26401 Harper
St. Clair Shores, MI 48081
313/779-8700

Automation Systems Inc.
208 N. 12th Ave.
Eldridge, IA 52748
319/285-8171

Bailey Controls Co.
29801 Euclid Ave.
Wickliffe, OH 44092
216/585-8500

Barber-Colman Co.
Industrial Instruments Division
P.O. Box 2940
Loves Park, IL 61132-2940
815/877-0241

Cincinnati Milacron Co.
Electronic Systems Division
Lebanon, OH 45036
513/494-5275

Control Technology Corporation
25 South St.
Hopkinton, MA 01748
617/435-9595

Crouzet Controls, Inc.
1083 State Parkway
Schaumburg, IL 60195
312/843-2200

Divelbiss Corporation
9776 Mt. Gilead Rd.
Fredericktown, OH 43019
614/694-9015

Dynage Controls
2 Willowbrook Rd.
Cromwell, CT 06416
203/635-6257

Eagle Signal Controls Div.
Wickes Mfg. Co.
8004 Cameron Rd.
Austin, TX 78753
512/837-8300

Eaton Corporation
Cutler-Hammer Products
4201 N. 27th St.
Milwaukee, WI 53216
414/449-6000

Electromatic Controls Corp.
2495 Pembroke Ave.
Hoffman Estates, IL 60195
312/882-5757

Encoder Products Co.
P.O. Box 1548
Sandpoint, ID 83864
208/263-8541

Entertron Industries
3857 Orangeport Rd.
Gasport, NY 14067
716/772-7216

The Foxboro Co.
420 Neponset Ave.
Foxboro, MA 02035
617/543-8750

Furnas Electric Co.
1000 McKee St.
Batavia, IL 60510
312/879-6000

GEC Automation Projects
27301 West Eleven Mile Rd.
Southfield, MI 48034
313/353-4800

GE Fanuc Automation
North America, Inc.
P.O. Box 8106
Charlottesville, VA 22906
804/978-5000

General Numeric Corp.
390 Kent Ave.
Elk Grove Village, IL 60007
312/640-1595

Giddings & Lewis Electronics Company
666 S. Military Rd.
P.O. Box 1658
Fond du Lac, WI 54935-7258
414/921-7100

Gould Incorporated
Industrial Automation Division
P.O. Box 3083
Andover, MA 01810
617/475-4700

Guardian Electric Manufacturing Co.
1550 W. Carroll Ave.
Chicago, IL 60607
312/243-1100

Honeywell Inc.
IPC Division
435 W. Philadelphia St.
York, PA 17315
717/848-1151

IDEC Systems and Controls Corp.
1213 Elko Dr.
Sunnyvale, CA 94089-2211
408/747-0550

Industrial Indexing Systems
626 Fishers Run
Victor, NY 14564
716/924-9181

Joucomatic Controls, Inc.
8107-S Arrowridge Blvd.
Charlotte, NC 28210-5676
704/527-4622

Kaye Instruments Inc.
15 DeAngelo Dr.
Bedford, MA 01730
617/275-0300

Klockner-Moeller Corp.
4 Strathmore Rd.
Natick, MA 01760
617/655-1910

Lehigh Fluid Power
Route 179
P.O. Box 248
Lambertville, NJ 08530-0248
609/397-3487

MTS Systems Corporation
P.O. Box 24012
Minneapolis, MN 55424
612/937-4000

Maxitron Corp.
21 Tamal Vista Blvd.
Suite 200
Corte Madera, CA 94925
415/924-0700

McGill Manufacturing Co.
Electrical Division
1002 N. Campbell St.
Valparaiso, IN 46383
219/465-2200

Minarik Electric Co.
224 E. Third
Los Angeles, CA 90013
213/624-3161

Mitsubishi Electric
Industrial Automation Division
800 Biermann Ct.
Mt. Prospect, IL 60056
312/298-9223

Modular Computer Systems, Inc.
1650 West McNab Rd.
Fort Lauderdale, FL 33310
305/974-1380

Navcom, Inc.
350 N. Main St.
Huron, OH 44839
419/433-7626

Omron Electronics, Inc.
Control Components Division
One E. Commercial Dr.
Schaumburg, IL 60195
312/843-7900

Phoenix Digital Corp.
2315 N. 35th Ave.
Phoenix, AZ 85009
602/278-3591

Polymont Corp.
P.O. Box 276
Dearborn, MI 48121
313/271-6990

Reliance Electric Company
Control Systems Division
4900 Lewis
Stone Mountain, GA 30083
404/938-4888

Siemens Energy & Automation, Inc.
Programmable Controls Division
10 Technology Dr.
Centennial Pk.
Peabody, MA 01960
617/532-6720

Solid Controls, Inc.
6925 Washington Ave. S.
Minneapolis, MN 55435
612/941-6110

Square D Co.
Automation Products Division
P.O. Box 472
Milwaukee, WI 53201-0472
414/332-2000

Struthers-Dunn Inc.
Systems Division
P.O. Box 1327
Bettendorf, IA 52722
319/359-7501

Telemecanique Inc.
901 Baltimore Blvd.
Westminster, MD 21157
301/876-2214

Tenor Co. Inc.
17020 W. Rogers Dr.
New Berlin, WI 53151
414/782-3800

Texas Instruments
Industrial Controls Department
P.O. Drawer 1255
Johnson City, TN 37601
615/461-2500

Texas Instruments
6918 Sierra Ct.
Dublin, CA 94568
415/829-6600

Thesaurus, Inc.
3322 S. Memorial Parkway Suite 205
Huntsville, AL 35801
205/880-0108

Toshiba/Houston
13131 West Little York Rd.
Houston, TX 77041
713/466-0277

Triconix Corporation
16800 Aston St.
Irvine, CA 92714
714/261-0880

Veeder-Root Co.
Digital Products Division
70 Sargeant St.
Hartford, CT 06102
203/527-7201

Westinghouse Electric Co.
Programmable Controls Department
200 Beta Dr.
O'Hara Township
Pittsburgh, PA 15238
412/963-4000

APPENDIX B

Adatek State Language

(Reprinted with Permission of Adatek, Inc.)

To: The Controls Engineering Community
From: Ray Pelland, President, ADATEK, INC.
Subject: The State Language Revolution

I'm always surprised at the difficulty in describing a new idea. I'm also surprised at the unwillingness of some engineers to explore and accept new and better approaches to problem solving. That's often the problem I face when talking to control engineers about State Language programming. It's just so much easier to hang on to that with which we are already familiar—like Ladder Diagrams.

If you will invest a few minutes in reading the attached paper, I guarantee you'll discover a way to make your job easier; more important, you will dramatically improve the quality of your work.

Programming languages are the tools of the control engineer, just like the tools in a carpenter's tool chest. As time goes by, all tools get better. Hand drills became power drills, slide-rules became calculators, relays became programmable controllers, Ladder Diagrams are becoming State Languages. Tools multiply and apply the talent of the control engineer.

The problem facing the control engineer is specifying and coordinating multiple, concurrent sequential tasks in a machine or process. PSM^{tm} has been created for that specific purpose. Its structure is that of the problem, a set of concurrent sequences.

It's difficult to describe all the advantages of this powerful design tool. First and most striking is the reduction in programming time. Most sequences can be described in a few minutes; major programs in a few hours. Then there's the ease of coordinating the machine's various sequential functions, something that seldom gets completely worked out in many Ladder Logic programs. With a State Language, the linkage between sequences is obvious, making synchronization much easier.

There are plenty of other advantages too. Like simplified troubleshooting at the machine level, or the ease of adding operator diagnostic messages. But the most important difference is the difference it makes to the control engineer and the quality of his work. Less time telling the computer what to do—more precision and elegance in the solution. The attached paper was written to give the reader a better understanding of the State Language revolution now underway.

IS A STATE LANGUAGE IN YOUR FUTURE?
THE EVOLUTION OF PROGRAMMABLE CONTROL

The programmable controller you specify in 1990 might look similar to the one you buy today. But under the cover, it's likely to have an entirely different personality. The need for an improved approach to control programming is causing a slow but steady evolution in PC system software.

The evolutionary path seems to be converging on some form of graphic flow chart language. The Europeans, via Siemens in Germany and Telemecanique in France have adapted such a language as a standard for programmable control. This language, Grafcet, uses sequential flow charts for programming. Grafcet is much like PSM (Process State Monitor), a language introduced by Adatek in 1982. Both Grafcet and PSM are languages in which the programmer directly defines the states that constitute the process activity.

Still, in the U.S., the ladder diagram remains dominant. What advantages do sequential flow chart languages have over the ladder? Will the ladder ever be replaced?

In many fields subject to computer automation, high order languages provide the shortest path between a problem and its solution. That's because a high order language is designed to model the characteristics of the particular type of problem to be solved. With an appropriate high order language, the programmer only needs to enter a description of the problem. With any computer language, the lower the language level, the more the programmer must be concerned with not only understanding and defining the problem, but telling the computer, sometimes in great detail, how to solve it.

An example of the effectiveness of high order languages is the popular spreadsheet program like Lotus 1.2.3.™ Modeled after the accountant's spreadsheet, the user only needs to describe the problem in terms of the values and the relationships between them. The program knows what to do to solve the "what if" question.

But is the spreadsheet actually a language? After all, you don't have to be a programmer to use it. That's just the point. A well designed high order language doesn't bear much resemblance to BASIC or other common computer languages. It will appear to the user as a convenient way to describe a particular kind of problem.

Interestingly, the ladder diagram is also a high order language. With it, the programmer enters a problem description. In this case the programmer describes an imaginary relay network that, if real, would cause the required system behavior. As a language model, the relay network was quite useful during the transition from relays to microcomputer based controls. However, it carries with it many of the disadvantages of real relay based control systems. The most serious of these is the inability to see the sequential activity of the process in the ladder program.

To come up with a better language model, it's first necessary to understand the kinds of problems that PCs solve. Essentially, PCs are concerned with the control of real physical devices with respect to time. This is fundamentally different from data processing computers whose main purpose is to process and manipulate pure information.

Time appears in the control program in two forms. One is sequential. A machine's parts cannot do two things at once. They must occupy various states, one at a time, and must obey the sequential rules dictated by the machine's design and the product's requirements. This is how a raw casting becomes a transmission case, through a carefully synchronized sequence of steps.

The complement to sequential activity in control is concurrency, or parallel activity. Production is increased if the separate parts of the machine can run their tasks simultaneously. This way a machine can include a loader, a punch and an ejector, working at the same time, processing three parts instead of one.

It appears then, that controlling physical devices with respect to time, both sequentially and in parallel, is the essence of programmable control.

Examining current control languages in this light shows why programming with today's today's PCs can be a frustrating experience.

The ladder diagram is primarily a parallel logic model. Sequential activity can be created with the ladder diagram by using a logical trick, the latching contact. But the more sequential the problem, the more permissive, preventive and latching contacts must appear on each rung, and the more confusing the program becomes. In most real applications it is impossible to simply look at the ladder diagram program and visualize how the machine is supposed to operate. This is why days or weeks can be consumed making simple changes to ladder diagram programs.

Data processing languages such as BASIC are "in-line" languages. While they tend to be slow, they can be used to specify sequential tasks. As long as they can execute their single sequential program without concurrently needing to know if the emergency stop switch is closed, they are adequate. But ask them to handle a few parallel alarms and the problem soon becomes overwhelming.

Even though control oriented data processing languages are now available that provide a level of concurrency, these languages still have a serious problem when it comes to control. Data processing languages do not scan.

```
*******************************************************
*                                                     *
*         Definition of Abbreviations for all Diagrams *
*                                                     *
*    CR - Control Relay    LT - Light         RDY - Ready   *
*                                                     *
*    ET- Elapsed Time      MAN - Manual        RET - Return  *
*                                                     *
*    FWD - Forward         PB - Push Button     SEC - Second *
*                                                     *
*    LS - Limit Switch     PIP - Part In Position SOL - Solenoid *
*                                                     *
*******************************************************
```

The Scanning PC Continuously Refreshes Its I/O

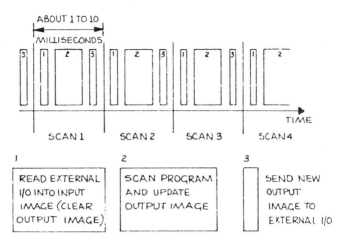

In addition to providing a framework for concurrency, scanning offers a more reliable environment for the control of potentially dangerous and expensive physical equipment. The scanning PC can correct a transient induced I/O error before it has any effect on the system. With the data processing language, or non-scanning controller, the error becomes permanent.

What is really needed for control is a language that permits the simple specification of both parallel and sequential process activity. Within that framework it should be able to interface to all types of physical processes through both digital and analog transducers. To handle a wide variety of applications, it should also provide data processing, like math and message communications. For maximum reliability the language should scan. PSM, the state language in Adatek's System 10 programmable controller, exemplifies such a language.

A state language PC is based on the idea that the controlled activity of a process can be effectively described with one or more state diagrams. Each diagram is essentially a flow chart that defines one sequential element of the process. Together, the set of state diagrams can describe almost any process. Typical control applications consist of one or more sequential elements such as motion control sequences, mode control sequences, motor start sequences, alarm sequences, etc.

Once the parallel sequences are entered, the state language PC makes them happen. If the programmer has described the activity accurately, the system will work.

It sounds simple, but in practice, describing a machine's activity is not trivial. It requires the ability to first step back from the process and consider the whole system as a set of sequential elements. Then the programmer must provide the detailed definitions for each state.

This is quite the opposite of ladder diagram programming where each output is placed on the right side of a rung, and then contacts and function boxes are added until the machine behaves. Although the state language approach will seem foreign at first to an experienced ladder diagram programmer, the payoff for approaching the control problem this way is considerable. The result is a structured, easy to modify program, in which the process activity is obviously visible in the program statements.

Reading a state diagram is simple. It's only necessary to know a few simple rules:

* Each state (a box and the arrows going out of it) in the diagram defines what will happen while that state is active.

* One state in each diagram will always be active.

* When power is first applied to the system, state number one in each sequence will automatically be the active state. From that point on, other states in the diagram can take their turn becoming the active state, but only according to the transition rules (the arrows) established by the programmer.

The Classic Start/Stop Circuit and Its State Diagram Equivalent

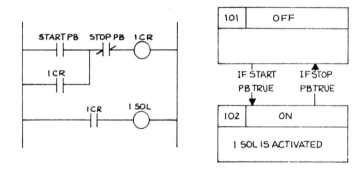

It is important to realize that a state diagram language is a high order language. Just as the Lotus 1.2.3.™ user does not tell the computer what steps to take to update the spreadsheet, a state language program is not written as a set of instructions for the computer to follow. Instead, it is a description of the problem.

For example, if the application requires that the FWD SOL is on while the MOTION sequence is in the FORWARD state, it's only necessary for the programmer to indicate that the FWD SOL will be ON in that state definition. There is no need to turn on the output when that state becomes active, nor turn it off when it becomes inactive. The state language PC takes care of those details.

To illustrate the power of this approach, consider the following machine composed of two sequential elements, a mode control sequence and a motion control sequence.

Remember that each arrow is a possible transition path. If an arrow's conditions become true while the state (that the arrow comes from) is active, that sequence will change to the new state. The actual transition does not consume any time. Therefore, if an output is defined as ON in two adjacent states, the output will not go OFF momentarily during the transition.

State Diagrams for a Typical Transfer Machine

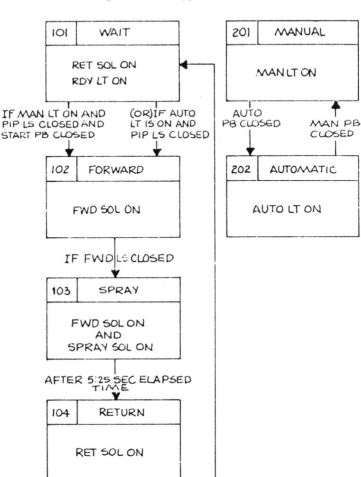

One of the main advantages to state language programming is that it only takes a brief explanation for non-programmers to read a state diagram. This way other interested folks can "walk through" the diagrams to see if that is really how the system is to behave, before the system is implemented.

With this powerful visual tool, potential problems can be addressed early in the design phase of the system. For example, a common problem with the type of machine described in the diagrams is the jammed transfer table. One strategy to minimize downtime due to jams would be to have the PC sense the jam and clear the machine. The modified diagrams for this system might then be:

Modified State Diagrams to Clear the Jam

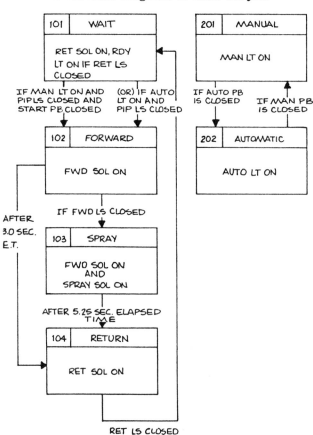

In this case an elapsed time test is used to specify an additional transition path from the FORWARD motion state. The new transition goes directly to the RETURN motion state, clearing the jam, but only if the table takes too long (3 SEC) to reach its intended destination, the FWD LS.

Notice that the programmer has also made the RDY LT ON function, in the WAIT state of the MOTION sequence, conditional upon the RET LS being closed. This way, if power is applied to the machine while the table is not at the RET position, the RDY LT will not come on until the table returns.

Other changes may be made as the programmer explores possibilities to increase the effectiveness and value of the system. With the state language, the emphasis lies in accurately describing, with the set of state diagrams, how the system is to work. Once the diagrams are done, they are simply entered into the controller in the appropriate form.

The details of the entry process can vary depending on which state language PC is being used. In the Adatek System 10, the PSM language has its own line based text editor. This way any serial ASCII terminal (or microcomputer running communications software) can be used to edit or download the program code. Others, like Grafcet, require a microcomputer development system of some kind. The advantage to this approach is that the state diagrams can be entered graphically, similar to the way ladder diagrams are edited and converted to PC usable code.

A state language is any language that permits the direct specification of multiple concurrent sequences. The power of the particular state language implementation will depend on the type of conditional and functional terms available to define process activity within that format.

The terms available in Adatek's PSM language can be used to define state activity for most possible control requirements. These include terms for analog and digital I/O, serial port communications, integer and floating point math, calls to user provided machine code (8088), PID loop specification and others.

If enough functional and conditional terms are available, a state language PC can provide an almost unlimited programming capability within the state format. For this reason, state language controllers are being applied to many projects beyond the capability of traditional ladder based PCs. Some of these include production testing applications that require long multiple sequences, data acquisition, report generation and interactive systems that require menu driven operator interfaces.

A perceived disadvantage of the state language is troubleshooting. How does an electrician approach a typical I/O related problem when the program is written in a state language format?

Once the method is understood, troubleshooting a state based control system becomes simpler than its ladder logic counterpart. That's because in the process of creating the state diagrams, the programmer has isolated and identified the particular I/O elements that are relevant for each state of the process.

For example, suppose that the transfer machine described here is stuck at the WAIT state of the motion sequence and that the mode sequence is in MANUAL. In AUTO mode the machine operates properly. First, looking at the ladder diagram would show several preventive, permissive and latching contacts directly or indirectly connected to the FWD sol.

The Partial Ladder Diagram Shows Several Possible I/O Faults

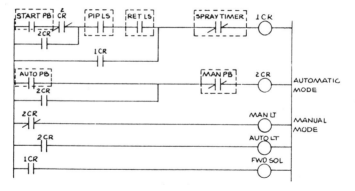

In this case a total of six I/O elements plus the SPRAY TIMER implementation (elsewhere in the program with its associated I/O) must be considered as potentially faulty circuits.

Conversely, the state diagram that describes the system is an ideal troubleshooting roadmap. This state diagram shows that when the MAN LT is ON, the motion sequence must switch to the FORWARD state when the START PB is actuated and the PIP LS is closed. In this case, the START PB and the PIP LS are the only two elements that can logically prevent the system from going into the FORWARD state.

The Partial State Diagram Isolates the Fault To One Circuit

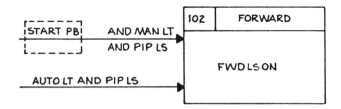

A clever troubleshooter would further notice that the PIP LS is also a required condition of the AUTO mode transition path, and since that one works OK, the problem can only be one place, in the START PB circuit.

The effectiveness of the state language is due in part to top down, structured programs that are easy to visualize and fast to troubleshoot. There are several other reasons why the state language is gaining in popularity. These can be summarized as:

* Control programming can undergo a change in emphasis. The programmer's focus can shift from implementation (how to arrange contacts and coils) to definition (what does the system do, and when does it do it?). Because the program is inherently structured and modular, it becomes easier to consider a variety of control strategies to make program changes and to visualize how the system will respond.

* The state diagrams that define a system become the control program. Because they come first, they can be effectively used to include management and other key people in the design process before the system is implemented. The system has a better chance to meet goals and stay within the budget.

* A state language PC is inherently faster than its relay logic counterpart. A state language PC only needs to scan about 10 to 30 percent of the program code that an equivalent ladder diagram PC would. This is because the state language program only scans the current active state of each sequence. In a ladder diagram PC, the contacts that affect the outputs can be anywhere in the program. Therefore, the ladder diagram PC must scan every rung.

* The state language program is almost self documenting. Instead of several sheets of ladder diagram drawings, most systems can be fully described with a few standard sized pages. This makes it less expensive to maintain the documentation and cuts down the time needed to find something in the program.

* Any physical system that operates through a set of sequential states can be described with the state diagram. This means that servo motion control, process (PID loop) control, work cell synchronization, batching systems, robotics and other special machinery can all be controlled with basically the same kind of program. The state language can become a standard for the control of any kind of physical system.

It's no wonder that once programmers begin using a state language, they find it difficult to consider any other method. The fact is that programmable control cannot remain limited to the constraints of a purely parallel logic model. The state model of multiple concurrent sequences provides the necessary ingredients to break those barriers.

Certainly no language can be more effective than one where the programmer simply describes the problem to the computer. With a state language, that's just how it works.

SYSTEM 10 USER'S GUIDE (11/4/85)
5.0 Introduction to State Language Programming

Most PLCs (programmable logic controllers) execute ladder diagram programs. The ladder diagram is an imaginary network of relay contacts and coils that if real, would cause the machine connected to it to behave in a particular way. Ladder diagrams are widely used because they are based on familiar technology and because they permit easy troubleshooting.

But is the ladder diagram the best possible language for control? Some wonder if BASIC, FORTH, PASCAL or some other language will ever replace the ladder diagram.

Ladder diagrams are complicated by one thing, the latching contact. By expanding the scope of the ladder diagram language into what is called a "state" language, latching contacts can be eliminated. Because the state language is a problem description language rather than an I/O specification language like ladder diagram, some users have already standardized on this approach. The state language is already in wide use and will see many new applications as more engineers become aware of this powerful tool.

5.1 Why Common Microcomputer Languages Don't Work

More about the state language later. First, let's focus on the difference between the ladder diagram language and common microcomputer languages like those mentioned. These general purpose languages provide a shorthand for the hundreds of machine code statements that would be required to do some simple task, like adding two numbers together. Even though they save a lot of program space, and in the case of PASCAL can be elegantly structured, the still operate the same way the machine code statements do. They tell the computer, step by step, how to solve a particular problem. Any language that can be used for many different purposes, like accounting, video games or statistical analysis has to be organized as a set of instructions for the computer to follow.

But programs written in the highest level languages are usually for more specific purpose. Because they are more specific, instead of telling the computer what to do, they can be written as a description of the problem to be solved. In a truly high level language, once the description has been entered, the computer knows what steps to take to solve it.

A good example of this difference is the typical spreadsheet program. The spreadsheet offers a problem description environment for the user. Only the mathematical relationships between the cells needs to be entered; the computer knows what to do with them. Doing matrix analysis with a general purpose language is certainly possible, but this can take hours compared with the few minutes required for the experienced spreadsheet user.

Even though the general purpose language is inefficient compared with a true high level language, the most important reason that it isn't suited for industrial control is that it doesn't scan. As simple as that sounds, the importance of scanning in a control system is often misunderstood.

5.2 The Importance of Scanning In Industrial Control

A production machine typically consists of several parts, like conveyors, indexers, control panels, etc. These parts must be coordinated with each other, but because they have different functions in the system they follow separate operating sequences. Scanning allows the PLC to handle these multiple concurrent (multitasking) requirements in an organized way.

As part of this process, scanning also provides the PLC with a snapshot, an instantaneous image of the machine's I/O status. By dealing with this snapshot instead of the actual I/O, the scanned PLC can eliminate "contact race".

Most important, in a scanning PLC, every few milliseconds, all the I/O in the system are rewritten to their correct states, even if no changes were required. In this way a scanned PLC can be virtually self healing in situations where high external noise levels might cause an I/O error.

The programmable controller has evolved over the last decade into a highly reliable control computer. Data processing microcomputers and control computers that execute data processing languages like BASIC don't scan. In short, the programmable controller is easier to use and it's a more practical and reliable choice for the control of industrial machinery and equipment.

5.3 The Ladder Diagram Obscures The State Definitions

A control program actually controls the system's states, the sequential events of the machine's operation. A state is simply a time when one particular thing is happening in the system. Anyone who has used relay logic already knows what a state is; they've intuitively designed lots of them into every ladder diagram.

Using relay logic, states are created with latching contacts. The classic START, STOP circuit is a two state system:

```
            START      STOP   CR1
        |----] [--+--]/[----O---|
        |         |             |
        |----] [--|             |
              CR1
```

At state 1 (the power-on state) CR1 is OFF. When the system is latched into state two, initiated by the START PB and held there by the CR1 latching contact, CR1 is ON. Subsequently depressing the normally closed STOP PB will unlatch the circuit allowing it to return to state 1, etc.

This same latching circuit can be used to illustrate the fundamental problem with the ladder diagram. Notice that the following version of the START, STOP circuit does essentially the same thing.

```
            START                   CR1
        |---] [----------+-----O----|
        |                |          |
        |---] [---]/[---|          |
             CR1   STOP
```

But what happens if the operator depresses both the START PB and the STOP PB at the same time? Now, each circuit will behave differently. In the first version the output CR1 will be held OFF, in the latter it will remain ON.

On this simple level, the response of the system to different contact arrangements is obvious. In a real system with several hundred contacts, coils and system states, predicting the system's response for all possible combinations can be nearly impossible.

The problem is that the ladder diagram controls the system's state changes indirectly, with contacts and coils distributed through the program. The system's response is "implied" from the arrangement. In a state language, the system's response is explicit, it is directly specified in the program.

5.4 The State Diagram

To illustrate the power of this concept, a state diagram of the same two state ON, OFF control system might look like this.

```
+-----------+                                      +-----------+
| STATE 1   |       START PB PUSHED                | STATE 2   |
|           |       AND STOP PB RELEASED           |           |
|           |     |-------------------------->|    |           |
|           |     |   STOP PB PUSHED             |  | CR1 IS ON|
|           |     |<--------------------------|    |           |
+-----------+                                      +-----------+
```

A state in a state diagram is a box and the arrows emanating from it. Notice that in the state diagram, what happens in each state of the system is defined. In this example the diagram explicitly states that the system will only change from state one to state two if the START PB is pushed while the STOP PB is released.

In the state diagram there is no possibility of misunderstanding either what happens at each state or the rules for changing from one state to the other. The state language program is simply a coded form of the state diagram that defines the system.

Before going further, it is important to understand that assigning "states" to a system is a subjective process. A state is a human invention, and how we assign them to a particular machine is somewhat arbitrary. In fact the power of state language programming is largely derived from the ability to assign states in such a way that key system outputs are accounted for in separate sequential states that closely follow the intended operation.

But designing a machine and its associated controls is not necessarily an easy task, no matter what kind of control system or language is used. We often learn a great deal about the system as we attempt to define it in detail. The state language offers a very flexible, fast development tool to aid in this learning process. With the state language you can easily reconsider the system definition in many different variations until the exact definition is attained.

5.5 A State Language Example

The Adatek System 10 exemplifies a programmable controller that executes state language programs. The System 10 program is a simple code of ASCII alphanumeric characters. A state language PLC could be designed to accept a program of graphic symbols. Adatek chose the ASCII code for System 10 so that any RS232 terminal can be used as a programming device.

To illustrate state language programming on the System 10, let's reconsider the state diagram shown earlier and assume that,

The START PB is connected to I/O channel # 101, and
the STOP PB is connected to I/O channel # 102, and
CR1 is connected to I/O #108,

. . . then state one of our two state system would be entered as follows,

101:T101N102G2

In the System 10, each state in the diagram is given a line in the program. The line number identifies the state. Line 101 is state one of the first sequence, 102 is state two of the first sequence, line 250 is the 50th state of the second sequence, etc.

Each line or state definition can contain several statements. Each statement begins with a colon. Each statement defines one of the things that will happen when this state is active. As you can see this state definition only has one statement. It says "(T101) if contact 101 is (T)rue" and "(N102) if contact 102 is (N)ot true", then "(G2) Go to state 2". Notice that this is the same as the state diagram box for state 1.

Or, reading the whole line, and referring to the actual I/O devices instead of their channel numbers, a narrative description of what is going on while this state is active can be read from the program code,

```
                                                           101:T101N102G2
                                                            |  |   |   |
While state one of sequence one is active ----------+  |   |   |
                                                        |   |   |
              if the START PB is closed --------------+   |   |
                                                            |   |
       (and) if the STOP PB is OPEN ------------------+   |
                                                                |
          (then) Go to state 2 ----------------------+
```

. . . and similarly for state two,

```
                                                     102:A108:T102G1
                                                      |  |    |   |
                                                      |  |    |   |
        While state two is active ---------+  |    |   |
                                            '     |    |   |
               (A)actuate CR1 -------------+    |   |
                                                      |   |
  (in addition) if the STOP PB is closed ------------------+   |
                                                                  |
          (G)o back to state 1 -----------------------+
```

Remember that the state language is a problem description language. Each state definition is a description of what is supposed to happen at a particular time, it is not a set of instructions for the computer to follow. So "actuate" here does not literally mean turn on the output. In a state definition, actuate means that output will be ON while this state is active.

Conversely, instead of inserting preventive contacts to keep an output OFF when it is not needed, in a state language program you simply change to another state, one that doesn't actuate that output.

To remain consistent with the definition of a state diagram, the states occupy time, the state changes do not. A sequence can change from one state to another instantaneously. If an output is defined as on in two different states, it will not go off momentarily as the sequence changes from one of these states to the other.

5.6 Typical Machine Control Application

To better see the practical side of state language programming, let's look at a typical machine control system. Because the state diagram is such a good descriptive tool, consider the following example:

Sequence 1 (Mode Control)

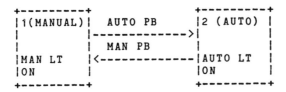

```
        +---------+                    +---------+
        |1(MANUAL)|  AUTO PB           |2 (AUTO) |
        |         |--------------->|   |         |
        |         |   MAN PB           |         |
        |MAN LT   |<---------------|AUTO LT   |
        |ON       |                    |ON       |
        +---------+                    +---------+
```

Sequence 2 (Motion Control)

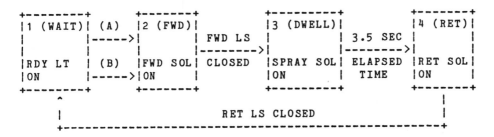

```
+---------+        +--------+            +----------+          +-------+
|1 (WAIT)|  (A)   |2 (FWD)|            |3 (DWELL)|          |4 (RET)|
|        |------->|        |  FWD LS    |          | 3.5 SEC |       |
|        |        |        |--------->|   |          |--------->|       |
|RDY LT  |  (B)   |FWD SOL| CLOSED     |SPRAY SOL| ELAPSED |RET SOL|
|ON      |----->|ON      |            |ON       | TIME    |ON     |
+---------+        +--------+            +----------+          +-------+
     ^                                                              |
     |                          RET LS CLOSED                       |
     +------------------------------------------------------------+
```

(A) = IF PART_IN_POSITION SW IS CLOSED AND AUTO LT IS ON, GO TO
       STATE 2.
(B) = IF START PB IS CLOSED AND MAN LT IS ON, GO TO STATE 2.

Like most machines, this system uses more than one state diagram to completely describe the machine's operation. Real systems can often be thought of as several small machines, each associated with a particular aspect of the systems total operation. In this case we have a mode control machine primarily associated with the operator's control panel and a motion control machine that controls the machine's motion actuators.

Having more than one sequence in a system is similar to having the finish carpenters, the electricians, and the plumbers working at the same time to complete a job. Although each crew has their own schedule or sequence to complete, they must be coordinated by the general contractor so that each task will mesh with the others at key points in the overall job.

The state language permits the programmer to act as the general contractor for the system. Using the state language he can specify the sequence of states for each major aspect of the process activity and in addition can specify important points in the sequences where they become synchronized with each other.

In all state language programs, one state in each sequence will always be active. The controller will start each sequence at state number one when power is applied. After that it's up to the statements in each state definition to determine when the other state changes will occur.

The two sequences in the diagram are independent for the most part, but the outputs of the mode control sequence are used as input information for the other. Note that the change from the WAIT state to the FWD state in the motion control sequence is controlled by the status of the mode control sequence.

Relay Logic and State Logic are Scanned the Same Way

There are actually two kinds of terms in a state language (or a relay logic) program; conditional terms and functional terms. The functional terms do something. The conditional terms don't do anything themselves, but if they aren't satisfied can keep all the terms that follow them in the statement from being executed.

This is the same way a ladder diagram rung is scanned. Conditional terms, contacts usually, precede functional terms, the coils, from left to right on each rung of the diagram. Each rung in a ladder diagram is equivalent to a statement in a state language program. In fact, if the G (state change) term is omitted, state language programs can become ladder diagram programs. Consider the following ladder rung.

```
|     106   107   108   112 |
|----] [--] [--] [--]/[---0--|
|                            |
```

All three conditional terms must be satisfied before coil 12 will be actuated. The state language statement (started with a colon) is scanned the same as a rung in relay logic. The statement,

:T106T107N108A112

. . . would perform the same function as the ladder rung above.

Back to the Example

To review, the conditional terms used so far are T and N, for true and not true (open or closed) contact conditions. The functional terms used are A and G, for actuate and goto another state in that sequence.

In the programming example, once the state definitions have been created, which is really just a description of how the machine works, in state diagram format, the only remaining task is to assign I/O numbers that correspond to the System 10 channels. For this example we can use the following arbitrary assignments:

Channel   Device

101   START PB
102   RET LS
103   FWD LS
104   PIP (PART IN POSITION) SW

105	MAN PB
106	AUTO PB
107	RDY LT
108	FWD SOL
109	MAN LT
110	AUTO LT
111	SPRAY SOL
112	RET SOL

The state language program can then be written directly from the state diagrams and the I/O references.

```
101:A109:T106N105G2
102:A110:T105G1
201:A107:T109T101G2:T110T104G2
202:A108:T103G3
203:A111:E350G4
204:A112:T102G1
```

exactly follow the state diagrams that define the system.

5.7 How About Timers

The only term that this program uses that hasn't been discussed so far is the E term (in state 203), an elapsed time conditional.

In a state language program, timers are very easy. Because elapsed time is often part of a state's definition, each sequence is given a timer that is automatically restarted whenever that sequence changes state. The sequence timers therefore, always know the elapsed time for the current active state in each sequence.

When an E term appears in a statement, its value (in hundredths of a second) is simply tested against the timer's current elapsed time. Eventually, on some scan, they will be equal, and at that time (and on all future scans) the condition will be satisfied. Essentially, the E term becomes a contact that is satisfied when that amount of time has elapsed.

5.8 Easy Troubleshooting and Program Changes

Once you become familiar with the few state language terms, the system's operation, as specified in the state diagrams, becomes very visible in the program. For this reason, troubleshooting and program changes are greatly simplified.

In fact, if you look closely at the WAIT state of the motion control state diagram you may find something that was overlooked. What happens if, when power is applied to the system, the transfer table is not at the home position? If it were not, and the system was in AUTO mode, then the motion control sequence could hang at state one if the part was not in position.

By changing line 201 slightly, we can add a statement to bring the table to home position in case it isn't already there when power is applied. (:N102A112) Adding another conditional term to the statement that actuates the RDY LT will prevent it from lighting until the fixture is at home. (:T102A107)

The new state definition then becomes:

201:N102A112:T102A107:T109T101G2:T110T104G2

Now the complete definition for state one of the motion control sequence can be read as,

(While state one is active)

1) If the RET LS (102) is open, actuate the RET SOL (112).
2) If the RET LS (102) is closed, actuate the RDY LT (107).
3) If the MAN LT (109) is ON and the START PB (101) is closed, GO to state 2.
4) If the AUTO LT (110) is ON and the PIP SW (104) is closed, GO to state 2.

Although there are many ways the programmer could have dealt with this start up problem, this simple change illustrates an important part of state language programming. Any number of statements can be included in a state definition until it provides exactly what the system requires. But elaborating one state definition won't complicate the rest of the program. Because only one state in a sequence can be active at any time, the statements on one line have no effect on the others.

Another benefit of state language programming should be evident from this example. Since the state definitions account for all the time the system is running, once the state definitions have been entered, the program is done. There are no loose ends waiting to be discovered when power is applied to the machine.

5.9 Built In, Run Time Diagnostics

The ability to easily elaborate a state definition without affecting the rest of the program is often used in state language programs to build run time diagnostics right into the program.

For example, suppose that in this system, it is known that the transfer table occasionally gets stuck in the FWD motion state. If we know that the table never takes more than 2.5 seconds to reach the FWD LS when it works properly, we can simply add the statement,

. . . to line 202, the FWD motion state, which now becomes,

$$202:A108:T103G3:E300G15$$

. . . Now, as long as the fixture reaches the FWD LS (103) within 3 seconds (300 hundredths), the sequence will operate as it did previously, but if the table ever gets jammed, the sequence will change to state 15 at 3 seconds. The programmer can now define state 15 to ring a buzzer, print a message or possibly even have the machine attempt to recover from the jam in some way.

5.10 Control Beyond the Relay

A major problem associated with ladder diagram programming is the inclusion of data processing functions in the control program. In the ladder diagram PLC these functions are usually treated as "function boxes". The function box becomes active whenever the contacts in front of it on the rung, and all those throughout the diagram that relate to those contacts become true.

The use of these non-relay type terms in the state language program is simplified because their appearance in the program can always be associated with a particular state. A counter for instance can always be incremented at the last state of the motion sequence to keep track of parts produced.

System 10 includes a set of data handling and arithmetic operators that can be placed anywhere in a state definition sequence. These operators can execute mathematical expressions using integer or floating point math, can provide relational decision making and can perform a host of additional data oriented functions.

Again, the ability to easily specify when the data operations will occur can greatly simplify their use in the control program.

Another area that relay logic falls behind is in long multiple sequences that execute concurrently. This type of application is quite common in testing.

An automotive manufacturer in Michigan is a good example. Here a group has the responsibility for designing testers that automatically exercise and test automotive instrument clusters as they are produced.

To thoroughly test an instrument cluster requires subjecting it to several long sequences of electrical and mechanical actuation in various combinations. To speed up the test, some of the sequences are executed concurrently.

Previously, because these long sequential programs were beyond the capacity of typical ladder diagram PLCs, controls for these testers were implemented with custom designed microcomputers.

Programming and program changes were done in assembly language by outside systems houses. Now with the simplicity of state language programming, this test group writes and modifies their own test programs. Program changes which previously required 3 to 4 weeks, are now completed in house, in one day or less.

5.11 A Possible Standard

The state language offers the following benefits:

1) Control programming can undergo a change in emphasis. The programmer's focus can shift from implementation (how to arrange contacts and coils) to definition (what does the system do, and when does it do it?). It becomes easier to consider a variety of control strategies, to make program changes and to visualize how the system will respond.

2) The state diagrams that define a system become the control program. Because they are required anyway, they can be effectively used to include management and other key personnel in the design process before the system is implemented. The system has a better chance to meet goals without increasing costs.

3) A state language PLC is inherently faster than its relay logic counterpart. A state language PLC only needs to scan about 10 to 20 percent of the amount of program code that an equivalent ladder diagram PLC would. This is because the state language program only scans the current active state line for each sequence. In a ladder diagram PLC, the contacts that effect the outputs can be anywhere in the program. So the ladder diagram PLC must scan every rung.

4) The state language program is almost self documenting. Instead of several sheets of ladder diagram drawings, most systems can be fully described with a few $8\frac{1}{2} \times 11$ pages including state diagrams, I/O cross reference, the program itself and narrative descriptions. Note that the example program was fully listed with 6 program lines of about 16 characters each. With fewer pages to reference, reviewing the documentation is easier and preparation costs are greatly reduced. In addition, the state language program is inherently structured and modular, two features sorely lacking in relay logic.

5) Any physical system that operates through a set of sequential states can be described with the state diagram. This means that servo motion control, process (PID loop) control, work cell synchronization, batching systems, robotics and other special machinery can all be controlled with basically the same kind of program. The state language can become a unifying factor for all types of control.

Since the introduction of state language programming by Adatek in 1982, several hundred state language PLC systems have been installed. These cover most typical machine control applications including welding controls, furnace and oven controls, energy management, plastic injection molding, pick and place arms, conveyor controls, testing, and a host of others. In most cases the state language PLC was chosen because of the simplification it brought to the project. In other cases the control requirements were beyond what ladder diagram based PLCs could have handled.

5.12 Summary

The ladder diagram PLC is a practical tool for industrial control. It scans and it permits simple troubleshooting because of the left to right format of the ladder diagram language. Because of the reliability of the scanning operating system and its ability to handle multiple concurrent tasks, the ladder diagram PLC is a much better choice for industrial control than data oriented microcomputers.

The state language PLC shares these important features. But because the state language has a term that can directly cause a state change, the state language can describe something more than a relay network; it can describe the actual system that is being controlled. For this reason, the state language alternative may be as close as you can get to an ideal language for the control of industrial equipment and processes.

Appendix C

Boolean Algebra Summary

Operations

$$\times \;=\; \text{AND}$$
$$+ \;=\; \text{OR}$$
$$\overline{A} \;=\; \text{NOT } A$$

Laws

Commutative

$$A+B \;=\; B+A$$
$$AB \;=\; BA$$

Associative

$$A+(B+C) \;=\; (A+B)+C$$
$$A(BC) \;=\; (AB)C$$

Distributive

$$A(B+C) \;=\; AB+AC$$
$$A+BC \;=\; (A+B)\,(A+C)$$

Absorptive

$$A(A+B) \;=\; A$$

DeMorgan's

$$\overline{(A+B+C)} \;=\; \overline{A}\,\overline{B}\,\overline{C}$$
$$\overline{(ABC)} \;=\; \overline{A}+\overline{B}+\overline{C}$$

Identities

$$AA \;=\; A$$
$$X(1+A+B+C+\ldots) \;=\; X$$
$$\overline{\overline{A}} \;=\; A$$
$$\overline{\overline{(AB)}} \;=\; AB$$
$$\overline{\overline{(A+B)}} \;=\; A+B$$
$$AB+A\overline{B} \;=\; A$$
$$A+\overline{A}B \;=\; A+B$$
$$AB+AC+B\overline{C} \;=\; AC+B\overline{C}$$

Appendix D

ASCII Character Code

BINARY	OCTAL	DECIMAL	HEXADECIMAL	ASCII	REMARKS
0000000	000	000	00	NUL	NULL
					TAPE LEADER
					CONTROL SHIFT P
0000001	001	001	01	SOH	START OF HEADING
				SOM	START OF MESSAGE
					CONTROL A
0000010	002	002	02	STX	START OF TEXT
				EOA	END OF ADDRESS
					CONTROL B
0000011	003	003	03	ETX	END OF TEXT
				EOM	END OF MESSAGE
					CONTROL C
0000100	004	004	04	EOT	END OF TRANSMISSION
					END
					CONTROL D
0000101	005	005	05	ENQ	ENQUIRY
					WHO ARE YOU
					CONTROL E
0000110	006	006	06	ACK	ACKNOWLEDGE
					ARE YOU
					CONTROL F
0000111	007	007	07	BEL	RING BELL
					CONTROL G

BINARY	OCTAL	DECIMAL	HEXADECIMAL	ASCII	REMARKS
0001000	010	008	08	BS	BACKSPACE FORM EFFECTOR CONTROL H
0001001	011	009	09	HT	HORIZONTAL TAB TAB CONTROL I
0001010	012	010	0A	LF	LINE FEED NEW LINE CONTROL J
0001011	013	011	0B	VT	VERTICAL TAB CONTROL K
0001100	014	012	0C	FF	FORM FEED PAGE CONTROL L
0001101	015	013	0D	CR	CARRIAGE RETURN END OF LINE CONTROL M
0001110	016	014	0E	SO	SHIFT OUT RED COLORED RIBBON CONTROL N
0001111	017	015	0F	SI	SHIFT IN BLACK COLORED RIBBON CONTROL O
0010000	020	016	10	DLE	DATA LINK ESCAPE CONTROL P
0010001	021	017	11	DC1	DEVICE CONTROL 1 TRANSMITTER ON READER ON CONTROL Q
0010010	022	018	12	DC2	DEVICE CONTROL 2 TAPE PUNCH ON CONTROL R
0010011	023	019	13	DC3	DEVICE CONTROL 3 TRANSMITTER OFF READER OFF CONTROL S
0010100	024	020	14	DC4	DEVICE CONTROL 4 TAPE PUNCH OFF CONTROL T

BINARY	OCTAL	DECIMAL	HEXADECIMAL	ASCII	REMARKS
0010101	025	021	15	NAK	NEGATIVE ACKNOW-LEDGE ERROR CONTROL U
0010110	026	022	16	SYN	SYNCHRONOUS FILE SYNCHRONOUS IDLE CONTROL V
0010111	027	023	17	ETB	END OF TEXT BUFFER LOGICAL END MEDIUM CONTROL W
0011000	030	024	18	CAN	CANCEL CONTROL X
0011001	031	025	19	EM	END OF MEDIUM CONTROL Y
0011010	032	026	1A	SUB	SUBSTITUTE CONTROL Z
0011011	033	027	1B	ESC	ESCAPE PREFIX CONTROL SHIFT K
0011100	034	028	1C	FS	FILE SEPARATOR CONTROL SHIFT L
0011101	035	029	1D	GS	GROUP SEPARATOR CONTROL SHIFT M
0011110	036	030	1E	RS	RECORD SEPARATOR CONTROL SHIFT N
0011111	037	031	1F	US	UNIT SEPARATOR CONTROL SHIFT O
0100000	040	032	20	SP	SPACE BLANK
0100001	041	033	21	!	
0100010	042	034	22	"	
0100011	043	035	23	#	
0100100	044	036	24	$	
0100101	045	037	25	%	
0100110	046	038	26	&	
0100111	047	039	27	'	APOSTROPHE
0101000	050	040	28	(	
0101001	051	041	29	)	
0101010	052	042	2A	*	
0101011	053	043	2B	+	

BINARY	OCTAL	DECIMAL	HEXADECIMAL	ASCII	REMARKS
0101100	054	044	2C	,	COMMA
0101101	055	045	2D	-	
0101110	056	046	2E	.	PERIOD
0101111	057	047	2F	/	
0110000	060	048	30	0	NUMBER 0
0110001	061	049	31	1	NUMBER 1
0110010	062	050	32	2	
0110011	063	051	33	3	
0110100	064	052	34	4	
0110101	065	053	35	5	
0110110	066	054	36	6	
0110111	067	055	37	7	
0111000	070	056	38	8	
0111001	071	057	39	9	
0111010	072	058	3A	:	
0111011	073	059	3B	;	
0111100	074	060	3C	<	
0111101	075	061	3D	=	
0111110	076	062	3E	>	
0111111	077	063	3F	?	
1000000	100	064	40	@	
1000001	101	065	41	A	
1000010	102	066	42	B	
1000011	103	067	43	C	
1000100	104	068	44	D	
1000101	105	069	45	E	
1000110	106	070	46	F	
1000111	107	071	47	G	
1001000	110	072	48	H	
1001001	111	073	49	I	LETTER I
1001010	112	074	4A	J	
1001011	113	075	4B	K	
1001100	114	076	4C	L	
1001101	115	077	4D	M	
1001110	116	078	4E	N	
1001111	117	079	4F	O	LETTER O
1010000	120	080	50	P	
1010001	121	081	51	Q	
1010010	122	082	52	R	
1010011	123	083	53	S	

BINARY	OCTAL	DECIMAL	HEXADECIMAL	ASCII	REMARKS
1010100	124	084	54	T	
1010101	125	085	55	U	
1010110	126	086	56	V	
1010111	127	087	57	W	
1011000	130	088	58	X	
1011001	131	089	59	Y	
1011010	132	090	5A	Z	
1011011	133	091	5B	[	
1011100	134	092	5C	\	
1011101	135	093	5D	]	
1011110	136	094	5E	^	CIRCUMFLEX
1011111	137	095	5F	_	UNDERLINE
1100000	140	096	60	`	ACCENT GRAVE
1100001	141	097	61	a	
1100010	142	098	62	b	
1100011	143	099	63	c	
1100100	144	100	64	d	
1100101	145	101	65	e	
1100110	146	102	66	f	
1100111	147	103	67	g	
1101000	150	104	68	h	
1101001	151	105	69	i	
1101010	152	106	6A	j	
1101011	153	107	6B	k	
1101100	154	108	6C	l	LOWER CASE L
1101101	155	109	6D	m	
1101110	156	110	6E	n	
1101111	157	111	6F	o	
1110000	160	112	70	p	
1110001	161	113	71	q	
1110010	162	114	72	r	
1110011	163	115	73	s	
1110100	164	116	74	t	
1110101	165	117	75	u	
1110110	166	118	76	v	
1110111	167	119	77	w	
1111000	170	120	78	x	
1111001	171	121	79	y	
1111010	172	122	7A	z	
1111011	173	123	7B	{	

BINARY	OCTAL	DECIMAL	HEXADECIMAL	ASCII	REMARKS
1111100	174	124	7C	|	VERTICAL BAR
1111101	175	125	7D	}	
1111110	176	126	7E	~	TILDE
1111111	177	127	7F	DEL	DELETE
					RUBOUT

Note that binary codes 0000000 through 0011111 are control codes and are thus not printed. Binary codes 0100000 through 1111111 actually produce printed characters.

Appendix E

Summary of Relay Ladder and Boolean Programming Symbols

LADDER SYMBOLS

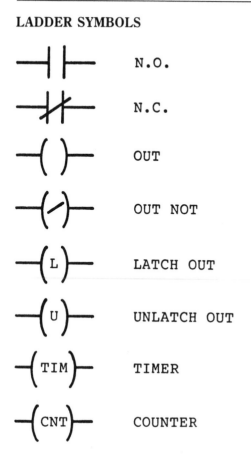

┤├	N.O.
┤╱├	N.C.
─( )─	OUT
─(╱)─	OUT NOT
─(L)─	LATCH OUT
─(U)─	UNLATCH OUT
─(TIM)─	TIMER
─(CNT)─	COUNTER

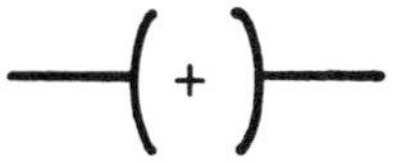

—(+)—   ADDITION

—(-)—   SUBTRACTION

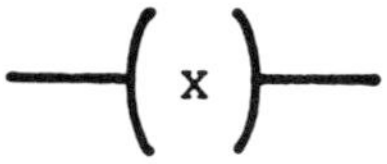

—(x)—   MULTIPLICATION

—(÷)—   DIVISION

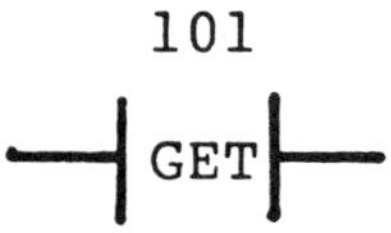

101
—| GET |—   GET INSTRUCTION

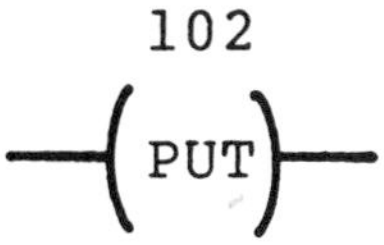

102
—(PUT)—   PUT INSTRUCTION

—| ↑ |—   OFF-TO-ON TRANSITIONAL ⎤ 
                                  ⎬   ONE-SHOT CONTACTS
—| ↓ |—   ON-TO-OFF TRANSITIONAL ⎦

—(MCR)—   MASTER CONTROL RELAY INSTRUCTION

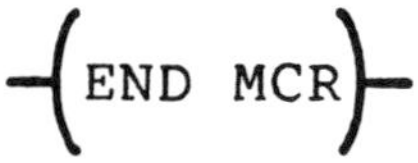

—(END MCR)—   END MCR INSTRUCTION

—(ZCL)—   ZONE CONTROL LAST INSTRUCTION

—(END ZCL)—   END ZCL INSTRUCTION

103
—(JMP)—   JUMP INSTRUCTION

104
—(JSB)—   JUMP-TO-SUBROUTINE INSTRUCTION

—(RET)—   RETURN-FROM-SUBROUTINE INSTRUCTION

105
—|CMP=|—   COMPARE EQUAL

106
—|CMP>|—   COMPARE GREATER THAN

107
—|CMP<|—   COMPARE LESS THAN

BOOLEAN SYMBOLS WITH LADDER EQUIVALENTS

BOOLEAN MNEMONIC   LADDER DIAGRAM

BOOLEAN MNEMONIC	LADDER DIAGRAM
AND	—| |—
OR	—| |—
OUT	—( )—
OUT NOT	—(/)—
NAND	—|/|—
NOR	—|/|—
LOAD	|—| |—
LOAD NOT	|—|/|—
OUT L	—(L)—
OUT U	—(U)—
TIM	—(TIM)—
CNT	—(CNT)—

BOOLEAN MNEMONIC	LADDER DIAGRAM
ADD	—(+)—
SUB	—(–)—
MUL	—(x)—
DIV	—(÷)—
CMP =	—(CMP =)—
CMP >	—(CMP >)—
CMP <	—(CMP <)—
JMP	—(JMP)—
JSB	—(JSB)—
MCR	—(MCR)—
END	—(END MCR)—

Appendix F
Selected
Supplier Literature

THIS APPENDIX IS DIVIDED INTO FOUR PARTS. PART 1 GIVES TECHNICAL DATA ON SMALL-to-medium PCs. Part 2 provides technical data on medium-sized PCs. Part 3 gives technical data on large PCs. Part 4 describes peripherals and accessories that are available with today's PCs, including local area networks. Readers who have completed the text of this book will find this appendix very informative, as it presents an opportunity to test their understanding of the material covered in the text.

Inclusion of suppliers' literature in this Appendix does not constitute an endorsement of the suppliers' products.

Literature on SY/MAX Products is reprinted courtesy of Square D Company. Literature on the Cutler-Hammer Product Line is reprinted with permission of Eaton Corporation, Cutler-Hammer Products. Data sheets on the Baby Bear Bones and Bear Bones PCs are reprinted with permission of Divelbiss Corporation, 9776 Mt. Gilead Rd., Fredericktown, OH 43019.

SMALL AND SMALL-TO-MEDIUM PCS

 Square D–Your Automation Foundation ™

TECHNICAL OVERVIEW

TO-101-01
JULY, 1986

SY/MAX ® **CLASS 8020 TYPE SCP-111, 112, 121, 122**
MODEL 100 PROGRAMMABLE CONTROLLER

The SY/MAX Model 100 Programmable Controller is a self-contained unit which includes the processor, Input/Output circuits, programmer interface, and power supply in a single compact package. No additional hardware is required with the exception of programming equipment.

The SY/MAX Model 100 Programmable Controller is housed in a rugged extruded aluminum case. Located on the front of the controller is a keyswitch used to select one of the three possible operating modes. Also included are five diagnostic LED (Light Emitting Diode) indicators used to indicate complete operational status of the controller as well as a communication port for connection to programming equipment and other devices.

The SY/MAX Model 100 Programmable Controller is available in two memory versions: RAM or UV PROM, and two I/O versions: 20 or 40 I/O. Terminal blocks on the front of the unit are the wiring points for I/O field devices such as pushbuttons, limit switches, motor starters and solenoids. The I/O section is used to convert high level signals (120 VAC 50/60 Hz) from the I/O field devices to the logic voltage level used by the processor section. Memory is housed in a removable module.

Power used to drive the processor logic, I/O circuitry, and programmer interface circuitry is obtained from an internal power supply. This supply converts the incoming voltage to the logic level used internally. Battery back-up is provided to retain program information stored in RAM memory, storage register data and latch relay status during power-out conditions. The battery is located on the memory module.

Mounting brackets are reversible for panel or flush mounting. Horizontal or vertical mounting is allowed.

SPECIFICATIONS

General

Ambient Temperature
Rating . . . . 0 to 60°C (32 to 140°F)
Operational
-40° to 80°C (-40 to 176°F) Storage
Humidity Rating . . . . 0-95% non-condensing

TYPE	DESCRIPTION	† APPROXIMATE WEIGHT (lb/kg)
Model 100 PC SCP-111,112	20 I/O PC	10.7/4.85
SCP-121,122	40 I/O PC	11.0/4.99

† Model 100 Approximate Weights (unpackaged)

Power Supply Section

Incoming Voltage Range . . . 94-132 VAC (47-63 Hz)

Input Power Consumption . . 25 VA*

Power Loss Ride Through
(at full rated load) . . . . 16ms (at 94 V)

Input Fuse Rating . . . . 3A 250 V

Input Fuse Type or
Equivalent . . . . BUSS ABC-3

*With either the Hand-Held Programmer or Loader/Monitor attached. Does not include output loading.

Battery Type . . . . One Class 8020 Type
SMM-115 Battery
(ETERNACELL No. TO4/42
3.4V Lithium Battery Size
1/2 AA or equivalent)

Battery Life - no load
(installed in PC with PC
power on) . . . . 3 years miminum at 60°C
Battery Life after ''BATTERY LOW'' LED Indicator
Illuminates. . . . . 2 weeks minimum
Battery Life - with load (processor RAM Memory Support Time) . . . . 1 year

MODEL 100 SPECIFICATIONS (Continued)

Processor Section

Logic: . . . . Microprocessor-based
Memory Size . . . . 420 Words (16 bit)
Memory Type . . . . RAM or UV PROM
Memory Utilization . . . . 1 word per contact, coil, branch
Memory Overhead . . . . None
Diagnostic Lights . . . . RUN, HALT, MEMORY, FORCE, BATTERY LOW
Scan Speed . . . . 40ms typical for a fully programmed memory
Instruction Set . . . . Relay Logic (Contacts, Coils)
Data Transfer
Latch/Unlatch Relays
Data Comparisons ($=, \neq, \geq, <$)
Timers (0.1 sec time base)
Counters (up and down)
Master Control Relay
Synchronous Shift Register
(Forward and reverse, 1, 8, or 16 channel)
Bit Read and Control
Transitional Output
I/O Forcing
Storage Registers . . . . 38 (4 digit)

I/O Section

External I/O . . . . 20 I/O - 12 Inputs/8 Outputs
or
40 I/O - 24 Inputs/16 Outputs
Internal Relay Equivalents
(Regular or Latching) . . . . . . up to 64

Inputs

Voltage Operation Range . . . 94-132 VAC (47-63 Hz)
Input Current Draw . . . . 11-16mA (60 Hz), 9-13mA (50 Hz)
Must Turn On Voltage 94V
Must Turn On Current . . . . . . 11.1mA (at 94V 60 Hz)
9.4mA (at 94V 50 Hz)
Must Turn Off Voltage 40V
Must Turn OFF Current . . . . 4.7mA (at 40V 60 Hz)
4.0mA (at 40V 50 Hz)
Input Impedance . . . . 8.5K ohms (at 60 Hz)
10.0K ohms (at 50 Hz)

Turn On/Off Time . . . . 8ms (60 Hz), 10ms (50 Hz) nominal plus scan time
16.6 (60 Hz), 20ms (50 Hz) worst case plus scan time
LED Operation Indication . . Red LED illuminates when the input receives an "ON" signal from the field input device
Compatibility with SY/MAX
Class 8030 Ouput Modules . . COM-221, DOM-221, GOM-221, DOM-225

Outputs

Voltage and Current Characteristics:

	VOLTAGE	CURRENT
Voltage Range	94-132V 47-63 Hz	—
Maximum Current	—	2 Amp/Output at 35°C (20 Amps total)*
Minimum Load	—	NONE
Maximum On State Voltage Drop Across Output	2.0 V at 2 Amp Load	—
Maximum Off State Leakage Current	—	2 mA at 132 V 50/60 Hz
Maximum Surge Current	—	25 Amp for 1 cycle (60 Hz) (Non-repetitive)

*1 Amp/Output at 60°C.

Turn On/Off Time . . . . 4.2 ms (60 Hz), 5ms (50 Hz) nominal**
8.3 ms (60 Hz), 10ms (50 Hz) worst case**
LED Operation Indication . . Red LED illuminated when receiving "ON" signal from processor logic
Compatibility with
SY/MAX Class 8030
Input Modules CIM-101, DIM-101

**Plus scan time (Note: Outputs turn on/off at AC sine wave zero crossing point.)

ORDERING INFORMATION

Class	Type	Description †
8020	SCP-111	Programmable Controller with 12 Inputs, 8 Outputs, and 420 Words RAM Memory
8020	SCP-121	Programmable Controller with 24 Inputs, 16 Outputs, and 420 Words RAM Memory
8020	SCP-112	Programmable Controller with 12 Inputs, 8 Outputs, and 420 Words UV PROM Memory
8020	SCP-122	Programmable Controller with 24 Inputs, 16 Outputs, and 420 Words UV PROM Memory
8020	SMM-100	Spare Model 100 RAM Memory Module
8020	SMM-110	Spare Model 100 UV PROM Memory Module
8020	SMM-115	Replacement Battery for use in Model 100 Programmable Controller
8020	SFK-210	Fuse Kit for Model 100, 10 Fuses
8020	SMB-120	Low Profile Mounting Bracket Set
8020	SMB-130	Horizontal Label Set

† All I/O are rated for 120 VAC operation and utilize the same power source.

For more information refer to Instruction Bulletin 30598-101-XX.

™
Square D–Your Automation Foundation™

TECHNICAL OVERVIEW

TO-103-01
JULY, 1986

SY/MAX ® CLASS 8020 TYPE SCP-311, 312, 313, 321, 322,
323, 332, 333, 344 SY/MAX MODEL 300 PROCESSOR

The SY/MAX Model 300 is a physically small yet very powerful processor designed for small to medium applications. It is enclosed in a sheet metal steel case, and contains processor logic and memory for storing the control program. The Model 300 is designed to plug into the processor slot of any SY/MAX digital or register rack. The processor also has five diagnostic LEDs and a block of diagnostic registers for ease of troubleshooting. To maintain system security the processor has a three position (RUN, HALT & DISABLE OUTPUTS) key operated mode selector switch.

The SY/MAX Model 300 Processor contains 112 (4 digit-16 BIT) registers. Each register can be assigned as either 16 digital I/O, 16 internal relays, 1 timer, 1 counter, or 1 data storage register. Intelligent SY/MAX I/O modules such as Analog and BCD I/O, High Speed Counter Input, Stepper Motor and Speech Output may be used with the Model 300. The recommended digital I/O capacity for a Model 300 processor with local I/O is 128 while the I/O capacity for a processor with remote I/O is 256.

The SY/MAX Model 300 is available in two functional versions: standard and deluxe. Standard processors perform relay ladder diagram logic, counting, and timing functions. In addition, the processor can be programmed for: 1) BCD to binary and binary to BCD conversions; 2) Master Control Relay; 3) synchronous shift registers; and 4) immediate I/O update. Processor functions can be programmed as transition sensitive functions. A single RS-422 compatible communication port allows communication between devices such as CRT programmers and other processors at distances up to 10,000 ft. (3600 m). The communication (BAUD) rate of this port is programmable to 9600 BAUD. Refer to Instruction Set table.

SY/MAX Model 300 Deluxe Processors have the following additional functions: 1) Asynchronous shift registers; 2) four function math (addition, subtraction, multiplication, and division, integer results to $\pm 32,767$); 3) ASCII output (PRINT); and 4) Communications. The Deluxe processor also features a second RS-422 compatible communication port. This port communication rate is also individually programmable to 9600 BAUD. Refer to the Instruction Set table.

The PRINT instruction allows alphanumeric messages to be generated from the ladder diagram program. These English language messages can be displayed on the CRT Programmer, Loader/Monitor, or on a printer. This makes the deluxe processor capable of alarm annunciation, report generation, and data logging with no special interfacing hardware.

The Communication instructions allow the processor to communicate directly (peer-to-peer) with other SY/MAX Family processors without requiring any special interfacing hardware. Only a direct cable connection is required for peer-to-peer communications. If a complete distributed control system is necessary, the Model 300 can be one of 200 SY/MAX devices that can communicate via the SY/NET® Local Area Network (LAN). Peer-to-peer communication allows I/O status and register data to be exchanged and enables outputs of one processor to be controlled by another processor. The SY/NET LAN provides peer-to-peer capabilities plus supports all programming functions over the network.

A special version of the Deluxe processor (SCP-344) is available with a RAM/UV PROM memory mix. This processor has a 2K word memory size with a memory combination of RAM and UV PROM. The UV PROM can be programmed to a maximum of 1K. The remaining unused memory can be programmed as RAM memory to a total size of 2K of both UV PROM and RAM. This allows critical circuits such as safety interlocks and emergency shutdown sequences to be programmed in UV PROM memory where they cannot be changed, while the RAM memory can be used for other functions which can be altered.

SPECIFICATIONS

Model 300 Processor

Environmental:
  Ambient Temp
  (Operational) . . . . 0° to 60°C
  Ambient Humidity . . . . . . 0 to 95% R.H.
  (non-condensing)
  Storage Temp. . . . . -40° thru +80°C
Dimensions (WxHxD):
  (Without Key) . . . . 1.5 x 12.8 x 6.6 inches
  (3.81 x 32.5 x 16.76 cm)
Weight (approximate): . . . . . 3.0 lbs./1.36 kg.
Rated Processor
Current . . . . SCP-311 . . . . 1500ma
Draw On SY/MAX   SCP-312 . . . . 1500ma
(+5V DC Output)   SCP-313 . . . . 1500ma
Power Supply   SCP-321 . . . . 1500ma
                    SCP-322 . . . . 1500ma
                    SCP-323 . . . . 1500ma
                    SCP-332 . . . . 2000ma
                    SCP-333 . . . . 2000ma
                    SCP-344 . . . . 2000ma
Capabilities:
  Logic. . . . . Microprocessor-based
  Memory Size. . . . . 1/2k, 1k, 2k (16 data BITS)
  Memory Type. . . . . RAM, UV PROM or RAM
  /UV PROM Mix
  Memory Utilization . . . . . 1 word per contact, coil,
  branch
  Memory Overhead . . . . . . 1/2K and 1K processors. . .
  6 words
  2 K processors . . . 128 words
  (Note: Type SCP-344 re-
  quires an additional 37
  words for the safeguarding
  feature.)
  Diagnostic Lights . . . . RUN, HALT, MEMORY,
  FORCE, I/O

Scan Speed. . . . . Approximately 30 ms/k
Addressing Capability:
Processor with Register & Remote I/O Capability:
  Total Addressable
  Registers . . . . 112 (4 digit-16 data BITS
  each)
  External I/O . . . . Over 250 (any mix in groups
  of 4 or 8
  Internal Relay
  Equivalents . . . . Over 250 (regular and latch-
  ing)
  Data Storage Registers . . Flexible to 112, 16 BIT data
  registers (may contain a 5
  digit signed decimal value
  from -32,768 to +32,767)
Processor with Local I/O Capability only:
  Total Addressable
  Registers . . . . 112 (4 digit-16 data BITS
  each)
  External I/O . . . . 128 max. (any mix in groups
  of 4 or 8)
  Internal Relay
  Equivalents . . . . 128 (regular and latching)

  Data Storage
  Registers . . . . Flexible to 112, 16 BIT data
  registers (may contain a 5
  digit signed decimal value
  from -32,768 to +32,767)

INSTRUCTION SET

INSTRUCTION	DISPLAY	DESCRIPTION	REMARKS
Contact	—⊣⊢—	Represents on/off status of an input, output, inter-nal relay, or register bit.	Input devices can be monitored in HALT.
Coil	—( )—	Represents an output connected to the processor.	Upon a transition to RUN, all coils are de-energized on the first scan. During the second scan, all coils are updated depending on the solu-tion of the logic in that rung.
Internal Relay	R- —( )—	Represents control relay function	Operation is much the same as an output but does **not** directly control output devices.
Latch/Unlatch Relay	—(L)— —(U)—	Duplicates the action of a mechanically held latch-ing relay. This is accomplished through the use of two internal relay coils with the same address. Once a latch coil is energized, it remains on until the unlatched coil is energized, even if power is removed.	Latched and unlatched rungs must be programmed sequentially. The second rung programmed will over-ride.
Transitional Coil	—(T)—	An internal or external coil which will only energize on a transition of the logic from open to closed. The output will remain on for a single processor scan. The logic must turn the coil off and transition to on again to re-energize the output coil.	Requires two consecutive I/O ad-dresses, one to record the transition and one to remember the last state. For this reason, a transitional coil must be assigned an ODD address number. The processor will then use the next I/O address for recording the last state of this same coil.

MODEL 300 INSTRUCTION SET (continued)

INSTRUCTION	DISPLAY	DESCRIPTION	REMARKS
Master Control Relay	MCR —( )—	MCR ON allows logic to function. MCR OFF disables outputs and logic functions. MCR controls all succeeding rungs until a new MCR coil is programmed or the end of program is reached.	
Timer	TMR	Timer register can store a four digit decimal value with a range from 0 to 9999. Selectable time bases are .01 sec., .1 sec., .1 min.	Programming techniques allow for programming of ON or OFF delay and interval timers. Replaces timing relays and motor driven timers.
Counter	CTR	Counter register can store a four digit decimal value with a range from 0 to 9999. UP, DOWN, and UP/DOWN counters are standard.	It is possible to cascade counters to obtain a count greater than four digits.
Synchronous Shift Register	SHFT	Used to shift individual bits of a storage register in either a forward or reverse direction. Possible to shift 1, 8, or 16 register BITS at a time.	Used in applications requiring sequencing and replaces drum switches.
Asynchronous Shift Register (First In, First Out)	FIFO	Data is loaded into the FIFO stack and exits out the last zone of the stack using IN and OUT commands respectively. 8 or 16 channel shifts are possible.	Used to record the order of events as they occur, such as in conveyor and alarm annunciation applications.
Data Comparison	IF (=, ≠, ≥, <)	Compares functions between storage registers or between a storage register and a constant. When an IF comparison is true, the IF box acts like a short.	It is possible to program 3 comparisons on one line. Capable of comparing a 5 digit number from -32,768 to 32,767. Register compares can support negative values. Math functions can be used in IF statements.
Data Transfer	LET	Transfer data from one storage register to another. Also performs math functions. Preset any constant value into a storage register (5 digit number from -32,768 to 32,767).	Occurs once per scan.
Transitional LET	TLET	Transition sensitive; restricts the operation of the LET instruction to one operation for each open to closed transition of the input condition.	
Binary to BCD	BCD	Conversion of binary data in a storage register to the binary coded decimal (BCD) equivalent of that data, within a LET or IF instruction.	Allows the processor to drive a BCD type LED display.
BCD to Binary	BIN	Conversion of binary coded data (BCD) in a storage register to the binary equivalent of that data. Used in a LET or IF instruction.	Used in applications to convert the on/off status from a thumbwheel switch to a binary number which the processor can operate on.
Four Function Math Addition Subtraction Multiplication Division	(+) (—) (x) (÷)	Math operations are performed left to right. Programmed in either IF or LET instructions.	
Immediate I/O Update	8176 —( )— -7	Setting bit 8176-07 will cause the processor to immediately update registers 1 thru 8 in the I/O image table memory.	
Forcing	CRT INITIATED	Forcing overrides the actual input status or outputs as controlled by the ladder diagram.	
Read	T READ	Transfers storage register and I/O data from remote device to local processor for peer-to-peer or LAN communication.	
Write	T WRTE	Transfers storage register and I/O data from a local processor to a remote device for peer-to-peer or LAN communication.	
Alarm	T ALRM	Transfers a numerical value to a register in a remote device. The number can range from -32,768 to 32,767.	
Print	T PRNT	Transfers ASCII coded messages from the processor to any device that communicates in ASCII.	Simplified means of producing alarm messages, production reports, etc.

TO-103-01

ORDERING INFORMATION

CLASS	TYPE	DESCRIPTION
8020	SCP-311	Standard Processor with 1/2K RAM Memory
8020	SCP-312	Standard Processor with 1K RAM Memory
8020	SCP-313	Standard Processor with 2K RAM Memory
8020	SCP-321	Deluxe Processor with 1/2K RAM Memory
8020	SCP-322	Deluxe Processor with 1K RAM Memory
8020	SCP-323	Deluxe Processor with 2K RAM Memory
8020	SCP-332	Deluxe Processor with 1K UV PROM Memory
8020	SCP-333	Deluxe Processor with 2K UV PROM Memory
8020	SCP-344	Deluxe Processor with 1K RAM/1K UV PROM Memory
8020	SLK-310	Model 300 Label Kit

For additional information refer to Instruction Bulletin 30598-103-XX.

1

CUTLER-HAMMER PROGRAMMABLE CONTROLLERS

D100 Micro Programmable Logic Controllers

T ECHNICAL
I NFORMATION
P UBLICATION
F200
(D100)

BASIC UNIT

The Basic Unit is a stand alone programmable logic controller containing I/O terminals, memory, and CPU (central processing unit). Basic Units are offered in three different sizes; 20 I/O, 40 I/O and 40 I/O High Performance. All units are equipped with the PROM Writer function so that EEPROM and UVPROM modules can be programmed in the basic unit from programs written on the programmer.

SPECIFICATIONS

Function	Controller Size		
	20 I/O	40 I/O ❶	40 I/O High Performance
Memory, words	512	1024	1024
Relays			
Internal	64	128	128
Retentive	64	128	128
Timers	8	16	64
Counters	8	16	64
Shift Registers	---	128	256
Flip-Flop	No	Yes	Yes
Step Sequencers	No	Yes	Yes
One-Shot	No	Yes	Yes
Master Control	Yes	Yes	Yes
Jump Coil	No	Yes	Yes
PROM Writer	Yes	Yes	Yes
Typical Scan Time	5 ms	10 ms	2.5 ms

❶ These functions also available on 20 I/O Controller with PG10A programmer.

EXPANSION UNIT

An Expansion Unit is connected by cable to the basic unit to provide additional I/O's. Two sizes are available — 20 I/O and 40 I/O.

PROGRAMMER

The Programmer will program all basic and expansion units in the D100 product family. The large LCD display allows 8 rungs, containing up to 10 elements, to be viewed at one time. The ladder logic programming is easy to understand and use. Start-up, diagnostics, and troubleshooting are also made simple because the ladder diagram will show "power flow" (contacts opening and closing, coils being energized and de-energized) while the D100 is running. The programmer comes complete with a cable.

DESIGN CHARACTERISTICS

The D100 Micro Programmable Logic Controller is a family of products for use in a wide spectrum of applications. The D100 Micto PLC is small in size and economically priced.

Features:
- Expandable 20 to 120 I/O
- Utilizes true ladder logic programming
- Programmer displays eight rungs of logic at one time
- Large LCD screen displays "power flow" while D100 is running

Power Supply	Input	Output Type	Output Volt	Catalog Number
20 I/O CONTROLLER (12 IN/8 OUT) W/MANUAL				
115 V ac	24 V dc	Relay	24 V dc – 230 V ac	D100CR20
115 V ac	24 V dc	Transistor	24 V dc	D100CD20
115 V ac	115 V ac	Relay	24 V dc – 230 V ac	D100CRA20
115 V ac	24 V dc	Triac	115 V ac	D100CA20
115 V ac	115 V ac	Triac	115 V ac	D100CAA20
24 V dc	24 V dc	Relay	24 V dc – 230 V ac	D100DCR20
40 I/O CONTROLLER (24 IN/16 OUT) W/MANUAL				
115 V ac	24 V dc	Relay	24 V dc – 230 V ac	D100CR40
115 V ac	24 V dc	Transistor	24 V dc	D100CD40
115 V ac	115 V ac	Relay	24 V dc – 230 V ac	D100CRA40
115 V ac	24 V dc	Triac	115 V ac	D100CA40
115 V ac	115 V ac	Triac	115 V ac	D100CAA40
24 V dc	24 V dc	Relay	24 V dc – 230 V ac	D100DCR40
40 I/O HIGH SPEED CONTROLLER (24 IN/16 OUT) W/MANUAL				
115 V ac	24 V dc	Relay	24 V dc – 230 V ac	D100CR40H
115 V ac	24 V dc	Transistor	24 V dc	D100CD40H
115 V ac	24 V dc	Triac	115 V ac	D100CA40H
115 V ac	115 V ac	Relay	24 V dc – 230 V ac	D100CRA40H
115 V ac	115 V ac	Triac	115 V ac	D100CAA40H
8 I/O EXPANSION UNIT W/30 cm CABLE 8 IN, 8 OUT OR 4 IN/4 OUT				
115 V ac	24 V dc	---	---	D100ENC8
115 V ac	115 V ac	---	---	D100ENA8
115 V ac	---	Relay	24 V dc – 230 V ac	D100ERN8
115 V ac	24 V dc	Relay	24 V dc – 230 V ac	D100ERC8
115 V ac	115 V ac	Relay	24 V dc – 230 V ac	D100ERA8
115 V ac	---	Triac	115 V ac	D100EAN8
115 V ac	---	Transistor	24 V dc	D100EDN8
20 I/O EXPANSION UNIT (12 IN/8 OUT) w/30 cm CABLE				
115 V ac	24 V dc	Relay	24 V dc – 230 V ac	D100ER20
115 V ac	24 V dc	Transistor	24 V dc	D100ED20
115 V ac	115 V ac	Relay	24 V dc – 230 V ac	D100ERA20
115 V ac	24 V dc	Triac	115 V ac	D100EA20
115 V ac	115 V ac	Triac	115 V ac	D100EAA20
24 V dc	24 V dc	Relay	24 V dc – 230 V ac	D100DER20
40 I/O EXPANSION UNIT (24 IN/16 OUT) w/30 cm CABLE				
115 V ac	24 V dc	Relay	24 V dc – 230 V ac	D100ER40
115 V ac	24 V dc	Transistor	24 V dc	D100ED40
115 V ac	115 V ac	Relay	24 V dc – 230 V ac	D100ERA40
115 V ac	24 V dc	Triac	115 V ac	D100EA40
115 V ac	115 V ac	Triac	115 V ac	D100EAA40

2

T ECHNICAL
I NFORMATION
P UBLICATION
F200
(D100)

CUTLER-HAMMER PROGRAMMABLE CONTROLLERS

7/1/86

D100 Micro Programmable Logic Controllers

Description	Catalog Number
MISCELLANEOUS	
LCD Programmer w/Cable	D100PG10
Printer Interface Module w/Parallel Port Cable	D100PF10
Timer Counter Access Module	D100 TCM10
EEPROM Memory Cartridge	D100EE10
UV PROM Memory Cartridge	D100UV10
Computer Interface Module	D100CIM16
Computer Compatible Software — allows programming, Troubleshooting, Monitoring and Documentation	D100CCS50
Computer Compatible Software Demonstration Disk	D100CCS10D
Spare Cable for Programmer (Cable Only)	D100AC10
Spare Cable for Expansion Unit — 30 cm (11.8")	D100AU30
Spare Cable for Expansion Unit — 50 cm (19.7")	D100AU50
Spare Operator's Manual (Pub. No. NC-185)	D100DM10
Spare Battery (Lithium — 3 V, 120 mAH)	D100AB10
Spare Fuse (250 V, 3 A)	D100AF10

APPROXIMATE DIMENSIONS
Dual Dimensions — Inches (mm)

Do not use for construction.

PRINTER INTERFACE MODULE

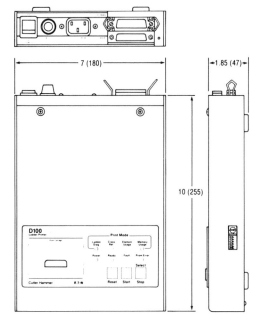

Power Requirements

20 I/O Controller . . . . 24 V dc, 20 W
. . . . 115 V ac, 50/60 Hz, 20 VA
. . . . 230 V ac, 50/60 Hz, 20 VA

40 I/O Controller . . . . 24 V dc, 25 W
. . . . 115 V ac, 50/60 Hz, 25 VA
. . . . 230 V ac, 50/60 Hz, 25 VA

Environmental Requirements

Controller . . . . 0°C–60°C operating
Programmer . . . . 0°C–40°C operating
Relative Humidity . . . . 20%–90% noncondensing

20 I/O CONTROLLER AND EXPANDER
APPROX. WEIGHT 2 lb. 8 oz. (1.1 kg)

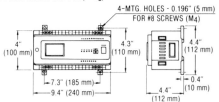

40 I/O CONTROLLER AND EXPANDER
APPROX. WEIGHT 3 lb. 6 oz. (1.5 kg)

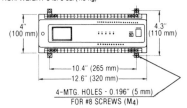

8 I/O EXPANDER
APPROX. WEIGHT 1 lb. 2 oz. (0.5 kg)

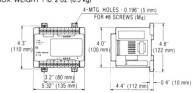

LCD PROGRAMMER OR TIMER COUNTER ACCESS MODULE
APPROX. WEIGHT 14 oz. (0.4 kg)

See note on installation and use of product at bottom of page 1.

E·T·N

2

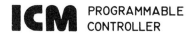 PROGRAMMABLE
CONTROLLER

BABY BEAR BONES PLC

DESCRIPTON

The Baby Bear Bones is a 5 input/5 output Programmable Logic Controller (PLC)
for the smaller system control applications. It's single board design is sized
to fit into industry standard brake-away plastic relay track or mount with
standard Bear Bones hardware kits. For more inputs and outputs, an 8 in / 8 out
expander is available. It's cable connected to the Baby Bear Bones. The inputs
are self supplying - no external voltage is needed. The outputs are normally
open relay contacts or an adjustable time delay module.

FEATURES

Internal time base (used for software timers)
On board logic supply
Self-supplied inputs
"Watch dog" led indicates the processor is running Visual indication of on/off
state of inputs and outputs. Outputs reset on loss of PLC primary voltage A.C.
and D.C. models.
96 hour burn-in at the factory
Compact and rugged

APPLICATION

This stand alone controller requires only a memory prom (with your program)
and connections to the real world. It is all you need to accept contact
closures and drive your solenoids or pick-up your motor starters. The inputs
and outputs that are available are 1/03 thru 1/07. The expander ICM-IO-20 adds
the capabilities of inputs and outputs 1/08 thru 1/15 see data sheet 7809-44.

The Baby Bear Bones will directly interface with one unit of other Bear Bones
line products i.e. solid state I/O expanders, TTL expanders, real time clock,
high speed counter, analog timer, etc. If more I/O is required, the standard
Bear Bones line is recommended.

PROGRAMMING

Programming is accomplished by using the Divelbiss ICM-PR-05 programmer. It
uses a ladder diagram format with special functions such as drums, counters,
timers sequencers etc. It will also "burn" your program onto a EPROM for
on-line operation. See manual ICM-UM-05 for more details. For patrons with
IM-01/TRS-80 interfaces, programming is the same as standard Bear Bones.

	Rev. 01 10/30/86
SHEET <u>1</u> OF <u>9</u>	NUMBER 7809—43

PRINTED IN USA

 **PROGRAMMABLE**
CONTROLLER

DATA SHEET

ICM-BB-12, 13

STANDARD OPTIONS

Timer module - Data Sheet #7809-45
16 I/O Expander - Data Sheet #7809-44

To provide a quick disconnect capability for the I/O terminals add a 0 to the
base part number.

    Example: An ICM-BB-13 becomes an ICM-BB-130.

Hardware mounting kit. Spacers, nuts, screws, lockwashers.

ON BOARD OPTIONS

See figure 2 page 8 for the location of the option pads and how to configure
them to your needs.

OPTION 1 - EPROM   The board is configured for a 4K EPROM. It can be
reconfigured for a 2K EPROM.

OPTION 2 - EXPANSION The board is configured for 10 functions it can be
reconfigured for 32 functions. If this option is selected output I/O3 is
sacrificed.

OPTION 3 - LINE FREQ. This option allows you to reconfigure the board from
60 to 50 HZ. The software timers will still be accurate.

OPTION 4 - INPUT SOURCE The Baby Bear Bone will furnish 12 VDC to your inputs.
You may change this to 5VDC on board. You may also elect to furnish your own
DC source up to 24VDC.

OPTION 5 - I/O EXPANDER The Baby Bear Bones can accept I/O expanders from the
Bear Bones family. This option must be selected and the board to be connected
must be selected for page 1, see 7809-28 page 2.

SPECIFICATIONS

CPU TYPE Single bit processor, 15 code instruction set.

PROGRAM MEMORY 4K x 8 capacity, selectable to 2K x 8, single supply, EPROM or
                ROM

TEMPORARY MEMORY (designated as internal "CR")
                256 x 1 capacity normal, 512 x 1 using expansion option.

User Range - CR1 - CR235 without expansion option
           CR1 - CR491 with option
           Subtract 12 CR7s from top for each special function used. SCAN
CLOCK 200kHZ ± 20%

TIME BASE .1 sec. period/100 HZ, readable at input I/O2
           Accuracy: BB-12 + .9% at 25°
           BB-13 dependant on 60HZ/50HZ line
           frequency

ᵞᵞ▐Divelbiss Rev. 01
CORPORATION 10/30/86

| SHEET 2 OF 9 | NUMBER 7809- 43 |

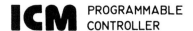

ICM PROGRAMMABLE
CONTROLLER

DATA SHEET
ICM-BB-12,13

SCAN TIME 5 micro sec./machine code instruction, 5 milli seconds for 1000
        instructions.

LOSS OF PRIMARY RESET Loss of primary to reset: 18mS
        duration: .2 sec min.

POWER SUPPLY 5VDC, ±5%, 1 Amp max.
        12 VDC, ±15%, .5 Amp max.

POWER CONSUMPTION Logic supply 1.2w max.
        I/O supply 2.2w max.
INPUT SUPPLY REQUIREMENTS BB-12 10.5VDC - 15VDC @ Amps max.
        BB-13 10.5VAC - 15VAC @ Amps max.

Note: Baby Bear Model BB-13 is supplied with step-down transformer.
      Input 90-130VAC @ 1 Amp max.

INPUTS Qty 5

		Sensitivity			
	Voltage	Turn On Resistance	Turn Off Resistance	Turn On Current	Turn Off Current
on board	5	1.5K	3.5K	.8mA	.3mA
selectable	12	10K	15K	.8mA	.3mA
user supply	24	25K	35K	.8mA	.3mA

	Response	
Voltage	Turn On	Turn Off
+5	2mS max.	15mS max.
+12	2mS max.	15mS max.
+24	2mS max.	2mS max.

Isolation: 1500V

Field termination: #14AWG max wire size per location. No lugs. Optional
quick-disconnect.

OUTPUTS Qty. up to 5

N.O. Relay contacts - isolated from each other
Relay module contacts rated 2 Amp at 120VAC or 28 VDC resistive.
Contacts are shunted with snubber circuitry to help surpress turn on
transcients.
FOR INDUCTIVE LOADS YOU MAY WISH TO CONNECT A VARISTOR ACROSS THE COIL OF THE LOAD.
   Varistor GE 130A0 or EQUAL.
OPTIONAL OUTPUTS

Adjustable time delay module(s). ICM-TM-05, 06, 07, 08
Ranges .1 to 6 sec or 1 to 120 seconds.
Refer to data sheet #7809-45.

LINE FUSE Customer supplied. Recommended .5 Amp, 220V no delay.

111Divelbiss CORPORATION

Rev. 01
10/30/86

SHEET	NUMBER
3 OF 9	7809- 43

 PROGRAMMABLE
CONTROLLER

DATA SHEET
ICM-BB-12,13

Operating Temp. Range: 0-60°C

Dimensions: 4.0 x 8.75", vertical clearance 2"

Weight: 0.4# less transformer

<u>WARNING</u>

The ICM Programmable Controller, as with other solid state controls, must not
be used in applications which would be hazardous to personnel in the event of
failure of the controller. Precautions must be taken to provide mechanical
and/or electrical safeguards external to the controller.

NOTE: Specifications subject to change without notice.

<u>LINE NOISE</u>

The Baby Bear Bones is designed to operate in normal industrial environments.
Should you experience problems with your service, please consider the following
suggestions.

<u>OUTPUT PROTECTION</u>

FOR INDUCTIVE LOADS YOU MAY WISH TO CONNECT A VARISTOR ACROSS THE COIL OF THE LOAD.

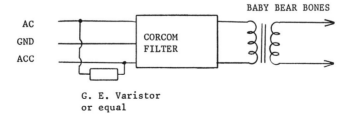

BABY BEAR BONES

AC
GND
ACC

CORCOM
FILTER

G. E. Varistor
or equal

Divelbiss
CORPORATION

Rev. 01
10/30/86

SHEET
4 OF 9

NUMBER
7809- 43

PRINTED IN USA

PROGRAMMABLE
CONTROLLER

DATA SHEET

ICM-BB-12, 13

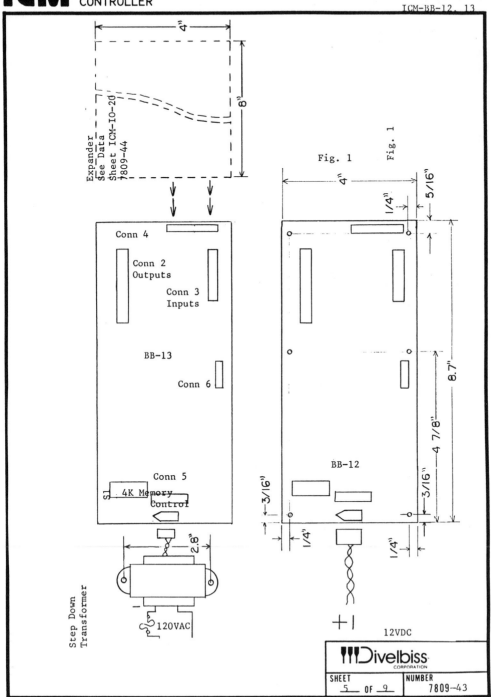

 PROGRAMMABLE
CONTROLLER

BABY BEAR BONES PLC

CONNECTOR 5

ICM programmers connected here can control the BABY BEAR BONES.

PIN

```
| 1  |
| 2  |
| 3  |
| 4  |
| 5  |
| C  |
| 7  |
| 8  |
| 9  |
| 10 |
```

1 Clock control
2 Connecting to ground halts the controller
3 Not used
4 Data Line
5 Not used
6 Not used
7 Not used
8 Keying pin, no connection
9 Preset input
10 Card common

CONNECTOR 6

Data buss accsess. Outputs are not buffered

PIN		FUNCTION
1		Reset out
2		System clock out
3		Word bit 3 out
4		Write bit out
5		Output data channel
6		Input data channel
7		Word bit 2 out
8		Word bit 1 out
9		Word bit $\emptyset$ out
10		Page bit 3 out
11		Page bit 2 out
12		Page bit 1 out
13		Page bit $\emptyset$ out
14		Card common

```
| 1  14 |
| 2  13 |
| 3  12 |
| 4  11 |
| 5  10 |
| 6   9 |
| 7   8 |
```

Note: If this connection is used,
       opt. 5 must be set. See
       Fig. 2, Page 8

!!!Divelbiss CORPORATION

SHEET	NUMBER
_6_ OF _9_	7809- 43

PROGRAMMABLE
CONTROLLER

DATA SHEET

ICM-BB-12, 13

CONNECTION 4 - Receptacle for BABY BEAR EXPANDER.

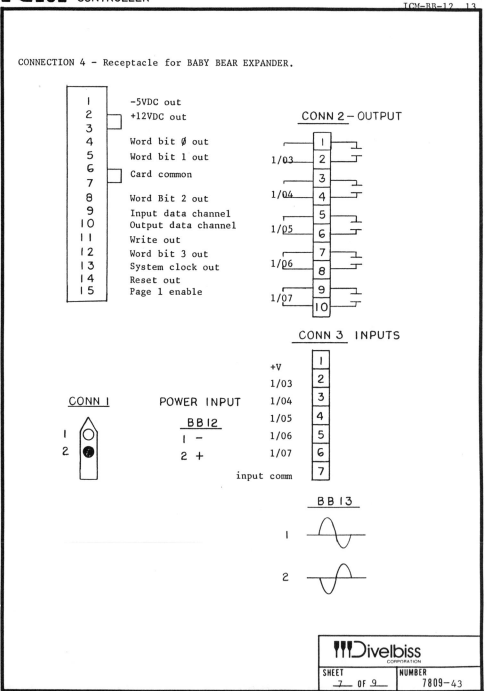

1	-5VDC out
2	+12VDC out
3	
4	Word bit Ø out
5	Word bit 1 out
6	Card common
7	
8	Word Bit 2 out
9	Input data channel
10	Output data channel
11	Write out
12	Word bit 3 out
13	System clock out
14	Reset out
15	Page 1 enable

CONN 2 - OUTPUT

CONN 3 INPUTS

CONN 1   POWER INPUT

BB 12

1 -
2 +

+V
1/03
1/04
1/05
1/06
1/07
input comm

BB 13

1
2

Divelbiss
CORPORATION

SHEET	NUMBER
7 OF 9	7809-43

PRINTED IN USA

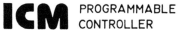 PROGRAMMABLE CONTROLLER

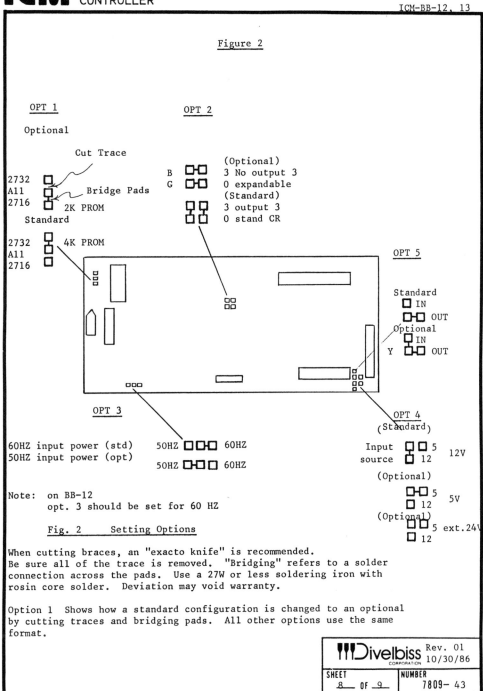

Figure 2

OPT 1

Optional

Cut Trace

2732
A11
2716

Bridge Pads

2K PROM

Standard

2732
A11
2716

4K PROM

OPT 2

(Optional)
B   3 No output 3
G   0 expandable
(Standard)
3 output 3
0 stand CR

OPT 5

Standard
IN
OUT
Optional
IN
Y   OUT

OPT 3

OPT 4
(Standard)

60HZ input power (std)
50HZ input power (opt)

50HZ 60HZ
50HZ 60HZ

Input   5   12V
source   12

(Optional)
5   5V
12

Note: on BB-12
opt. 3 should be set for 60 HZ

(Optional)
5 ext.24V
12

Fig. 2   Setting Options

When cutting braces, an "exacto knife" is recommended.
Be sure all of the trace is removed. "Bridging" refers to a solder
connection across the pads. Use a 27W or less soldering iron with
rosin core solder. Deviation may void warranty.

Option 1 Shows how a standard configuration is changed to an optional
by cutting traces and bridging pads. All other options use the same
format.

Divelbiss CORPORATION

Rev. 01
10/30/86

SHEET	NUMBER
8 OF 9	7809- 43

PRINTED IN USA

ICM PROGRAMMABLE CONTROLLER

DATA SHEET
ICM-BB-12, 13

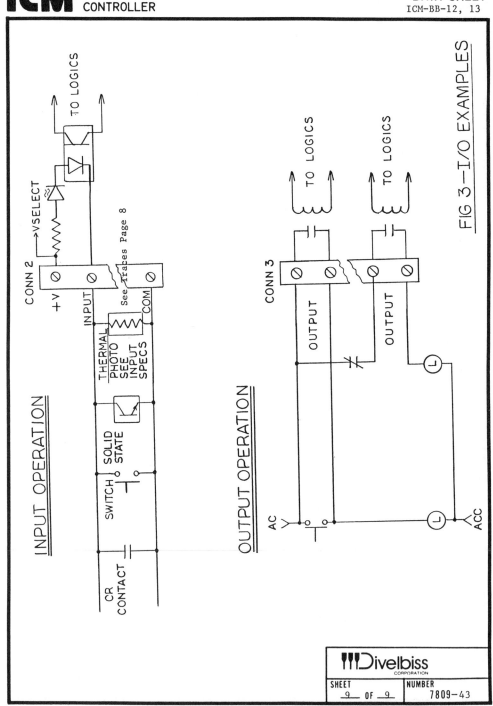

INPUT OPERATION

OUTPUT OPERATION

FIG 3—I/O EXAMPLES

Divelbiss CORPORATION

SHEET	NUMBER
9 OF 9	7809-43

PRINTED IN USA

PROGRAMMABLE
CONTROLLER

BEAR BONES CONTROLLER

DESCRIPTION

The ICM-BB-XX series of single board programmable controllers is a complete system, you need only add a programmed memory chip and your controller is ready. This single board contains all the hardware for real world inputs, real world outputs, and the power supply with the transformer mounted off the card.

TABLE 1

BEAR BONES (107 CR's)	BEAR BONES PLUS (491 CR's)	Line PWR	I/O GROUPS In	Out
ICM-BB-03	ICM-BB-30	120VAC	2	2
ICM-BB-04	ICM-BB-40	120VAC	1	1 & 3
ICM-BB-05	ICM-BB-50	120VAC	3	1 & 3
ICM-BB-06	ICM-BB-60	24VAC	3	1 & 3
ICM-BB-11	ICM-BB-110	240VAC	4	4

(See Table 2)

SPECIFICATIONS

CPU TYPE	Single bit processor
MEMORY	2K (ICM-ME-02) or 4K (ICM-ME-07)
INPUTS	See Table 2
OUTPUTS	See Table 2
POWER	120VAC at 25°C, 14.4W all 8 I/O OFF, 15.8W all 8 I/O ON
INSTRUCTIONS	15 (See manual ICM-OP-01)
SCAN TIME	5 usec per instruction
TEMPERATURE RANGE	0 to 60°C
DIMENSIONS	8"H x 9"W x 3"D (allow 2" on dimension H to mount transformer)
CLOCK	200KHZ
TIME BASE	0.1 second, I/O location 1/02
FIELD TERMINATIONS	14 AWG maximun wire size, with or without lugs.
WATCH DOG LED	Blinks to indicate that the system clock is operating.
LINE FUSE (TS-3)	Recommended 1 amp no delay
5VDC SUPPLY	Will operate entire system and will drive up to 40 outputs at one time. The outputs will require a connection to your real world power.

Divelbiss CORPORATION Rev. 05
01/01/87

SHEET	NUMBER
1 OF 8	7809- 26

| PROGRAMMABLE | DATA SHEET |
| CONTROLLER | ICM-BB-XX |

BEAR BONES CONTROLLER

APPLICATION

This stand alone controller requires only a memory prom (with your program) and connections to the real world. It is all you need to accept contact closures and drive your solenoids or pick-up your motor starters. This controller is dedicated to Page 1 of the I/O address set. The inputs and outputs that are available are 8 thru 15. The CUB expander ICM-IO-30 adds the capabilities of inputs and outputs 3 thru 7, see data sheet 7809-27. Should further expansion be required refer to ICM-IO-XX data sheet 7809-28 to add inputs and outputs in groups of 8 each. The maximum I/O count is 125 inputs and 125 outputs.

OPERATION

These controllers utilize the 14500 single bit processor. Instructions are read one at a time. The result of the instruction and the status of the data line are stored in the resultant register. If a normally open contact is programmed and the data line is high the resultant register is set to one. If a normally closed contact is programmed and the data line is low the resultant register is set to one. If a standard output symbol is programmed and the resultant register is one the output is energized. If a complimented output symbol is programmed and the resultant register is one the output is de-energized.

OPTIONS

The Bear Bones Plus offers additional features. There are 491 user control relays available. This product features break away terminal strips. This minimizes change out time and reduces the possibility of wiring errors since wires need not be disconnected.

Divelbiss CORPORATION Rev.05 01/01/87

| SHEET | NUMBER |
| 2 OF 8 | 7809-26 |

PRINTED IN USA

PROGRAMMABLE CONTROLLER

DATA SHEET
ICM-BB-XX

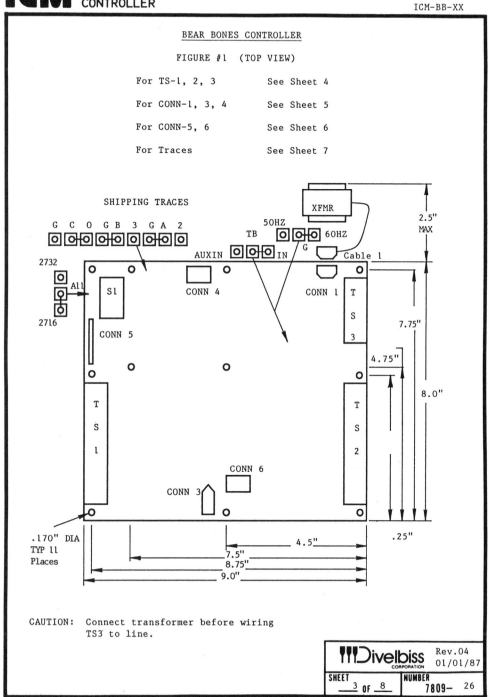

BEAR BONES CONTROLLER

FIGURE #1 (TOP VIEW)

For TS-1, 2, 3   See Sheet 4

For CONN-1, 3, 4   See Sheet 5

For CONN-5, 6   See Sheet 6

For Traces   See Sheet 7

SHIPPING TRACES

G C O G B 3 G A 2

50HZ
TB   60HZ
AUXIN   IN
G
Cable 1

2732
All
2716

S1
CONN 4
CONN 1

XFMR

2.5" MAX

T S 3

7.75"

4.75"

8.0"

CONN 5

T S 1

T S 2

CONN 6

CONN 3

.170" DIA
TYP 11
Places

.25"

4.5"
7.5"
8.75"
9.0"

CAUTION: Connect transformer before wiring
TS3 to line.

Divelbiss CORPORATION

Rev.04
01/01/87

SHEET
3 OF 8

NUMBER
7809- 26

 **PROGRAMMABLE
CONTROLLER**

DATA SHEET
ICM-BB-XX

BEAR BONES CONTROLLER

TERMINAL STRIPS

    Interfaces line power, input devices, and output devices to the BEAR BONES and BEAR BONES PLUS.

TS 3

GND
ACC
AC

The Ground wire terminates here.
The Grounded conductor terminates here.
The Ungrounded conductor terminates here.

TS 1

08	Input 08
09	Input 09
10	Input 10
11	Input 11
C	Common to Above
C	Common to Below
12	Input 12
13	Input 13
14	Input 14
15	Input 15

EXAMPLE OF EXTERNAL
CONNECTIONS FOR INPUTS

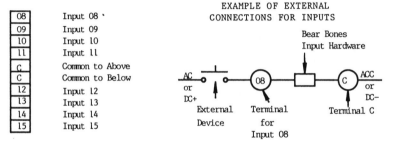

TS 2

08	Output 08
09	Output 09
10	Output 10
11	Output 11
C	Common to Above
C	Common to Below
12	Output 12
13	Output 13
14	Output 14
15	Output 15

EXAMPLE OF EXTERNAL
CONNECTIONS FOR OUTPUTS

NOTE: For output connections AC and ACC may be interchanged. See sheet 8 for additional connection comments.

 **Divelbiss** CORPORATION

Rev.05
01/01/87

SHEET	NUMBER
4 OF 8	7809- 26

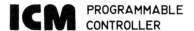 **PROGRAMMABLE**
CONTROLLER

DATA SHEET
ICM-BB-XX

BEAR BONES CONTROLLER

CONNECTOR 1

Interfaces the transformer to the BEAR BONES. Transformer mounted separate from board.

 Pins
4 & 6 Connects the transformer primary to TS 3
1 & 3 Connects the transformer secondary to the power supply
2 & 5 Connects the transformer secondary to the time base circuit.

CONNECTOR 3

  Pin
1   Card Ground

2   +5 VDC Logic Supply

3   Card Ground

CONNECTOR 4

Interfaces the CUB expander I/O to the BEAR BONES.

Pin
1   Card Ground
2   Not connected (see page 7, Clock Jumper)
3   Input #3
4   Input #4
5   Input #5
6   Input #6
7   Input #7
8   Card Ground
9   Output #3
10 Output #4
11 Output #5
12 Output #6
13 Output #7
14 +5 VDC Logic Supply

NOTE: Only 5 input and 5 output points are available with CONNECTOR 4. You can connect only one type of expander here. See 7809-27, 7809-32, 7809-35.

NOTE: The BEAR BONES PLUS does use output 3 for executive programming. This output is not available to you.

!!!Divelbiss Rev. 05
CORPORATION 01/01/87

SHEET	NUMBER
__5_ OF _8_	7809- 26

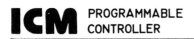 PROGRAMMABLE
CONTROLLER

DATA SHEET
ICM–BB–XX

BEAR BONES CONTROLLER

CONNECTOR 5

ICM programmers connected here can command the BEAR BONES.

Pin

1 Connecting to ground advances the program one step.

2 Connecting to ground halts the controller.

3 Not used.

4 +5 VDC = Data buss High; O VDC = Data Buss Low.

5 +5 VDC = Halted; O VDC = Running.

6 Page Bit 3 status; +5 VDC for pages 8-F; o VDC

7 +5 VDC = Result register is high

8 Keying pin, no connection

9 Connecting to +5 VDC resets outputs and resets program to step 000.

10 Card ground

11 Logic supply (+5 VDC)

12 Page Bit 2 status; +5 VDC for pages 4-7 and C-F; O VDC for pages 0-3 and 8-b.

13 Page Bit l status; +5 VDC for pages 2,3,6,7,A,b,E,F, O VDC for pages 1,4,5,8,9,C,d.

14 Page Bit 0 status; +5 VDC for pages 1,3,5,7,9,b,d,F, O VDC for pages 0,2,4,6,8,A,C,E.

CONNECTOR 6

Interfaces the I/O expanders to the BEAR BONES.

Pin

1 Connecting to +5 VDC resets outputs

2 Program clock synchronizes the I/O expanders and the BEAR BONES

3 IO/CR Bit 3 status

4 +5 VDC = write to outputs; O VDC = Read from inputs

5 Data Channel for outputs

6 Data Channel for inputs

7 IO/CR bit 2 status

8 IO/CR bit l status

9 IO/CR bit Ø status

10 Same as connector 5 pin 6

11 Same as connector 5 pin 12

12 Same as connector 5 pin 13

13 Same as connector 5 pin 14

14 Card Ground

Rev. 04
01/01/87

SHEET
6 OF 8

NUMBER
7809- 26

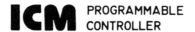 **PROGRAMMABLE CONTROLLER**

DATA SHEET
ICM-BB-XX

BEAR BONES CONTROLLER

CAUTION!   Be sure to remove 1/8" minimum of a trace when a cut is called for. Jumpers must be soldered. The user is responsible for his soldering techniques. Please feel free to consult ICM Applications at the Home Office.

50/60 HZ JUMPER

The BEAR BONES is shipped with the 60 HZ trace intact. To convert the 50 HZ cut the trace between G and 60 HZ, the install a jumper from G to 50 HZ.

2716/2732 EPROM JUMPER

The BEAR BONES is shipped with the 2716 pad connected to All. The BEAR BONES PLUS is shipped with the 2732 pad connected to All.

CLOCK JUMPER

The BEAR BONES is shipped with the IN (internal) trace intact. To convert to the AUX (external) capability cut the trace between TB and IN, then install a jumper from TB to AUX IN.

SHIPPING TRACES

The BEAR BONES is shipped with traces C to Ø, B to G, and A to G intact. The BEAR BONES PLUS is shipped with traces C to Ø, B to 3, and A to G intact.

Also contact the factory prior to attempting to make changes in the RAM jumpers, you may disable the programmable timers/counters.

THE REAL WORLD

While we design special circuitry including opto coupling, to isolate the controller from noise, external snubbing may be beneficial when energizing inductive loads.

W A R N I N G

The ICM Programmable Controller, as with other solid state controls, must not be used in applications which would be hazardous to personnel in the event of failure of the controller. Precautions must be taken to provide mechanical and/or electrical safeguards external to the controller.

NOTE: Specifications subject to change without notice

!!!Divelbiss CORPORATION	Rev.05 01/01/87

SHEET	NUMBER
7 OF 8	7809- 26

ICM PROGRAMMABLE CONTROLLER

DATA SHEET
ICM-BB-XX

BEAR BONES CONTROLLER

INPUTS

GROUP	SIGNAL LEVEL	POWER	FUSE	TURN ON/OFF	ISOLATION	LED	OPTO
1	90-130VAC	1.2W	1 AMP	25MS MAX	1500V	Y	Y
2	7-32VDC	1.3W	1 AMP	10/25 MS MAX	1500V	Y	Y
3	10-40VAC	1.2W	1 AMP	10/25 MS MAX	1500V	Y	Y
4	90-260VAC	2.4W	1 AMP	25MS MAX	1500V	Y	Y

NOTE: FOR GROUPS 1,3,4 CONNECT ACC TO COMMON. FOR GROUP 2 CONNECT DC- TO COMMON.

OUTPUTS

GROUP	SIGNAL LEVEL	POWER	FUSE	TURN ON/OFF	OVER VOLT.	0 X	ISOLATION	LED	OPTO
1&3	12-130VAC;.1-2A	240W	2AMP	1/2 CYCLE	400 PEAK	Y	7500V	Y	Y
2	7-32VDC; 2A	64W	2AMP	5 MS MAX	80VDC	N/A	1500V	Y	Y
4	35-260VAC;.1-1A	240W	1AMP	1/2 CYCLE	600 PEAK	Y	7500V	Y	Y

NOTE: FOR GROUPS 1,3,4 CONNECT AC TO COMMON TO AVOID SWITCHING THE NEUTRAL CONDUCTOR.
FOR GROUP 2 CONNECT DC- TO COMMON.

INPUT/OUTPUT SPECIFICATIONS

ΨΨΨDivelbiss CORPORATION   Rev.06   01/01/87

SHEET	NUMBER
8 OF 8	7809- 26

PRINTED IN USA

MEDIUM-SIZED PCS

Square D–Your Automation Foundation™

TECHNICAL OVERVIEW

TO-105-01
JULY, 1986

SY/MAX® CLASS 8020 TYPE SCP-521, 522, 523, 544
MODEL 500 PROCESSOR

The SY/MAX Model 500 is a physically small yet very powerful processor designed for medium applications. It is enclosed in a sheet metal case and contains processor logic and memory for storing the user's control program. The Model 500 is designed to plug into the processor and adjacent register slot of any SY/MAX register rack assembly. The processor has five diagnostic LEDs and a block of diagnostic registers for ease of troubleshooting. To maintain system security the processor has a three position (RUN, HALT & DISABLE OUTPUTS) key operated mode selector switch.

The Model 500 processor has the ability to address up to 2008 registers for the performance of its control functions. Each register can be assigned as a data value, counter, timer, 16 digital I/O, etc. Up to 460 of these registers may reside within the processor with the remainder residing within the SY/MAX Local Interface Modules (LI) and within other intelligent I/O modules such as Analog and BCD I/O, High Speed Counter Input, Stepper Motor and Speech Output Module.

The LI allows for serial communication between the I/O modules and CPU at distances up to 15,000 ft. (4572m). The local interface module registers can be used for external digital and intelligent I/O in addition to internal relays and data storage registers. Each local interface module has the capacity for 2048 digital I/O. Although the Model 500 processor can support seven local interface modules (14,000 I/O), a realistic I/O capacity is 2048, dependent on selected memory size, and application requirements. The ability to communicate to multiple LI modules results in reduced system throughput time.

The SY/MAX Model 500 Processor has all the functions of the Model 300 Deluxe Processor (ASCII Output, peer-to-peer or network communications, asychronous shift registers, two communication ports with programmable BAUD rates) with additional advanced capabilities such as 1) expanded math functions (addition, subtraction, multiplication, division, square root, random number, absolute value, etc., with integer results to $\pm 32,767$), 2) scan control (GOTO function, Subroutine and Timed Interrupt), 3) Matrix operations (block data functions), 4) Immediate Communications Update, 5) logical functions (AND, OR,

Exclusive OR), 6) program and communication software security features including program view inhibit, etc. Refer to the Instruction Set table.

The Model 500 processor is available in 2K, 4K and 8K RAM versions or a 8K RAM/UV PROM mix. The RAM/PROM mix allows for a maximum of 5461 words to be programmed in UV PROM. The remaining unused memory can be programmed as RAM memory to a total memory size of 8K. This allows critical circuits such as safety interlocks and emergency shutdown sequences to be programmed in the UV PROM memory where they cannot be changed, while the RAM memory can be used for other functions which can be altered.

Multi-level security features can be used to prevent the alteration of data and control programs, prevent I/O forcing, restrict network access of information, and provide password security.

NEW LISTING
Printed in U.S.A.

AJR

SPECIFICATIONS

Model 500 Processor

Rated Processor Current Draw On SY/MAX Power Supply
Type SCP-521 . . . . 4000ma
Type SCP-522 . . . . 4000ma
Type SCP-523 . . . . 4000ma
Type SCP-544 . . . . 4000ma
Environmental:
Ambient Temp
(Operational) . . . . 0° to 60°C
Ambient Humidity . . . . . . 0 to 95% R.H.
(Non-condensing)
Storage Temperature . . . . -40° thru +80°C
Dimensions (WxHxD):
(Without Key) . . . . 3.0 x 13.2 x 6.6 inches
(7.62 x 32.5 x 16.76 cm)
Weight (Approx.) . . . . 4.0 lbs./1.81 kg.
Scan Speed . . . . Approximately 2.6 ms/k (relay logic only)
Noise Testing . . . . Designed to meet NEMA ICS2-230
(Showering Arc) and ANSI C37.90a
(Surge Withstand Capability)

Capabilities:
Logic. . . . . Microprocessor-based
Memory Size. 2k, 4k, 8k
Memory Type. . . . . RAM or RAM/PROM
Memory Utilization . . . . . 1 word per contact, coil
Memory Overhead . . . . . . 47 words
Diagnostic Lights . . . . . . RUN, HALT, MEMORY, FORCE, I/O
Total Addressable
Registers . . . . Flexible to 2008, 16 Bit data registers (460 registers of which are processor resident).
External I/O. . . . . Over 2000 (any mix in groups of 4 or 8 I/O points per module)
Internal Relay
Equivalents . . . . Over 2,000
Data Storage Registers . . Flexible to 2008, 16 Bit data registers (may contain a 5 digit signed decimal value from -32,768 to +32,767)

INSTRUCTION SET

INSTRUCTION	DISPLAY	DESCRIPTION	REMARKS
Contacts	—∣—	Represents on/off status of inputs, outputs, internal relays, and register bits.	Input devices can be monitored in HALT.
Coil	—( )—	Represents an output connected to the processor.	Upon a transition to RUN, all coils are de-energized on the first scan. During the second scan, all coils are updated depending on the solution of the logic in that rung.
Internal Relays	R- —( )—	Represents control relay functions	Operation is much the same as an output but does not directly control output devices.
Latch/Unlatch Relays	—(L)— —(U)—	Duplicates the action of a mechanically held latching relay. This is accomplished through the use of two internal relay coils with the same address. Once a latch coil is energized, it remains on until the unlatch coil is energized, even if power is removed.	Latch and unlatch rungs must be programmed sequentially.
Transitional Coil	—(T)—	An internal or external coil which will only energize on a transition of the logic from open to closed. The output will remain on for a single processor scan. The logic must turn the coil off and transition to on again to re-energize the output coil.	Requires two consecutive I/O addresses, one to record the transition and one to remember the last state. A transitional coil must be assigned an ODD address number. The processor will then use the next I/O address for recording the last state of this same coil.
Master Control Relay	MCR —( )—	MCR ON allows logic to function. MCR OFF disables outputs. MCR controls all succeeding rungs until a new MCR coil is programmed or the end of program is reached.	

MODEL 500 INSTRUCTION SET (continued)

INSTRUCTION	DISPLAY	DESCRIPTION	REMARKS
Timer	TMR	Timer register can store a four digit decimal value with a range from 0 to 9999. Selectable time bases are .01 sec., .1 sec., .1 min.	Programming techniques allow for programming of ON or OFF delay and interval timers. Replaces timing relays and motor driven timers.
Counter	CTR	Counter register can store a four digit decimal value with a range from 0 to 9999. UP, DOWN, and UP/DOWN counters are standard.	Possible to cascade counters to obtain a count greater than four digits.
Synchronous Shift Register	SHFT	Used to shift individual bits of a storage register in either a forward or reverse direction. Possible to shift 1, 8, or 16 bit channels at a time.	Used in applications requiring sequencing and replaces drum switches.
Asynchronous Shift Register (First In, First Out)	FIFO	Data is loaded into the first zone and exits out the last zone using IN and OUT commands respectively. Possible to shift 8 or 16 bit channels at a time.	Used to record the order of events as they occur, such as in conveyor and alarm annunciation applications.
Data Comparison	IF $(=, \neq, \geq, <)$	Compare functions between storage registers or between an integer or floating point storage register and a constant. When an IF comparison is true, the IF box acts like a short.	It is possible to program 3 comparisons on the line. Integer mode is capable of comparing a 5 digit number from -32,768 to 32,767. Register compares can support negative values. Math functions can be used in IF statements.
Data Transfer	LET	Transfer data from one storage register to another. Also performs integer or floating point or combined mode math.	Ranges are same as Data Compare (above). Occurs once per scan.
Transitional LET, IF	T LET T IF	Transition sensitive; restricts the operation of the LET or IF instruction to one operation for each open to closed transition of the input condition.	
Indirect Register Read	RDSTAT RDDATA FNDBIT	Indirect register read of status field. Indirect register read of data field. Locate and clear the first bit (lowest bit number) set in the indirect register being read.	
Binary to BCD	BCD	Conversion of binary data in a storage register to the binary coded decimal (BCD) equivalent of that data, within a LET or IF instruction.	Allows the processor to drive a BCD-type LED display.
BCD to Binary	BIN	Conversion of binary coded data (BCD) in a storage register to the binary equivalent of that data. Used in a LET or IF instruction.	Used in applications to convert the on/off status from a thumbwheel switch to a binary number which the processor can operate on.
Standard Math Operations Addition Subtraction Multiplication Division Square Root Absolute Value Negate	LET, IF + — x ÷ SQRT ABS NEGATE	Math operations are performed left to right. Programmed in IF or LET instructions. Value range for integer math operations: -32,768 to +32,767 Change sign of current result.	Intermediate results to $\pm 2,147,483,647$ are valid.
Accumulator Manipulations	ACCALT ADDALT SUBALT DUPALT DUPACC SWAP 16 SWAP 32	Exchange current accumulator with the alternate accumulator. Add current intermediate result to the alternate accumulator. Subtract current intermediate result from the alternate accumulator. Copy alternate accumulator into current accumulator. Copy current accumulator (intermediate result) into alternate accumulator. Exchange low and high bytes of the current intermediate result. Exchange low and high words of the current intermediate result.	These manipulations allow the user to preserve 32-bit precision when performing calculations which extend beyond a single rung.
Identify Number Type	NUMTYP	Identifies mathematical classification of current intermediate result.	

TO-105-01

MODEL 500 INSTRUCTION SET (continued)

INSTRUCTION	DISPLAY	DESCRIPTION	REMARKS
Forcing	CRT INITIATED	Forcing overrides the actual input status or output as controlled by the ladder diagram.	
Serial communications to other processors or peripherals	T READ T WRTE T ALRM T PRNT	Transfer storage register and I/O status between devices which may be located up to 10,000 ft. apart. All are transition sensitive.	Requires no additional interface hardware.
Read	T READ	Transfers storage register and I/O data from remote device to local processor for peer-to-peer or LAN communication.	
Write	T WRTE	Transfers storage register and I/O data from a local processor to a remote device for peer-to-peer or LAN communication.	
Alarm	T ALRM	Transfers a numerical value from a local processor to a register in a remote device. Peer-to-peer or LAN. The number can range from -32,768 to 32,767.	
Print	T PRNT	Transfers ASCII coded messages from the processor to any device that communicates in ASCII.	Simplified means of producing alarm messages, production reports, etc. ASCII data format is selectable.
AND	$\wedge$	A logical function used within an IF or LET command which has the property such that if X and Y are two logic variables, then the function "X AND Y" is defined by the following table: X Y X AND Y 0 0 0 0 1 0 1 0 0 1 1 1	
OR	$\vee$	A logical function used within an IF or LET command which has the property such that if P and R are logic quantities, then the logical result of "P OR R" assumes values as defined in the following table: P R P OR R 0 0 0 0 1 1 1 0 1 1 1 1	
Exclusive OR	$\oplus$	A logical function used within an IF or LET command having the property such that if S and T are logic quantities, then the logical result of "S XOR T" is defined by the following table: S T S XOR T 0 0 0 0 1 1 1 0 1 1 1 0	
Random Number Generator	RANDOM	Generates random number.	
GOTO	GOTO	Instruction to cause the processor to skip over a section of the program to a designated position in the program. Position is determined by a MARK rung placed in memory.	Shortens processor scan time and immediately jumps to a designated position.

MODEL 500 INSTRUCTION SET (continued)

INSTRUCTION	DISPLAY	DESCRIPTION	REMARKS
Subroutines	GOSUB MARK ST SUB RTN	Group of ladder diagram rungs that can be executed from the user's ladder diagram only by enabling a GO SUB instruction which is programmed in the main program. Subroutines consist of a start of a subroutine (MARK), one or more ladder rungs, and a return (RTN). All subroutines must be programmed after the ST SUB instruction which denotes the beginning of the subroutine area. Nesting (subroutines within subroutines) are allowed up to 8 levels deep.	Applications include variable sequence machines, recipe programs, repeated programs.
Timed Interrupt, Communications Inhibit, Scan Time Limit	Reg. 8165	A special safeguard rung using register 8165 allows the following features: 1. Communications Inhibit 2. Program Viewing Inhibit 3. Safeguard Registers 4. Timed Interrupt-Scan Control 5. Scan Time Limit	Protect Programs used in conjunction with 8176 ⊢( )⊣ - 16 Scan Control
Matrix	M LET M IF M	A matrix is a group of storage registers handled as a unit. Formulates multi-register data transfers and math functions. Also allows multi-register compare statements.	Application includes machine diagnostics, menu selection, etc.
Immediate Communications Update	Reg. 8176 —( )— -11, 12, 13 and 14	Processes incoming data values from COMM port channels which were received during processor scanning.	

ORDERING INFORMATION

CLASS	TYPE	DESCRIPTION
8020	SCP-521	Model 500 Processor with 2K RAM Memory
8020	SCP-522	Model 500 Processor with 4K RAM Mmeory
8020	SCP-523	Model 500 Processor with 8K RAM Memory
8020	SCP-544	Model 500 Processor with 2.5K RAM/5.5K UV PROM

For additional information refer to Instruction Bulletin 30598-105-XX.

7/1/86
New TIP

CUTLER-HAMMER PROGRAMMABLE CONTROLLERS

MPC1 Modular Programmable Logic Controllers

T ECHNICAL
I NFORMATION
P UBLICATION
F200
(MPC1)
1

CONTROLLER

Digital and analog processors support 32 registers that can be used for counters, timers, shift registers, bit registers, and pointer registers. They can perform mathematical functions (add, subtract, greater than, less than, equal to, not equal), tasks normally reserved for much more costly PLC's.

The analog processor, in addition to the same functions as the discrete version, supports "intelligent" analog input and output modules and can perform double precision multiplication and division functions.

I/O modules plug into field wired, removable terminal blocks. Modules can be replaced in seconds. Entire PLC can be replaced without disconnecting wiring. Operates from 115 to 230 V ac. Internal battery back-up retains memory for three years without external power.

I/O EXPANDERS

Both processors can support up to three discrete I/O Expansion Chassis of 32 I/O each. The chassis can be mounted 2,000 ft. apart for an 8,000 ft. loop.

PROGRAMMER

The MPC1 Programmer will accommodate either the discrete or analog processor. It's LCD display guides the user through programming with self-prompting messages even advising when an error is made.

All programming is done in familiar, easily understood ladder logic — lengthy training sessions are not required.

The Programmer has a battery backed RAM memory and can be powered by the MPC1 or an optional desk top power supply. It can write programs to the PLC, storage cartridges, EEPROM chips or a printer. It can also read programs from the PLC, storage cartridges, or EEPROM chips.

DESIGN CHARACTERISTICS

A complete Programmable Logic Controller system for applications up to 128 I/O. Easily understood and operated . . . programs in relay ladder logic with digital and analog capabilities.

I/O MODULES

Durable, sealed I/O units provide exceptional noise immunity. Four I/O points per module with two points per common. Input module operates on ac or dc voltage. Ac and dc output modules are fused. All modules can be keyed and labeled for enhanced safety and identification. Unique lever handle makes for speedy replacement.

Analog I/O modules are intelligent peripherals that can perform mathematical operations on data independent of the processor, making the overall system faster and more efficient.

Operations such as scaling numbers to and from engineering units, input signal filtering, output signal conditioning, and linear and logarithmic ramping of the output voltage are examples of tasks performed by the modules.

Product Description	Catalog Number
DISCRETE (ONLY) COMPONENTS	
Controller (Supports 32 I/O points only)	MPC1C10
Controller (Supports 32 I/O points + exps.)	MPC1C20
I/O Expander Chassis (Supports 32 I/O points, ea.)	MPC1C30
Programmer	MPC1C10
ANALOG/DISCRETE COMPONENTS	
Controller (Supports 32 I/O points only)	MPC1C12
Controller (Supports 32 I/O points + exps.)	MPC1C22
Programmer	MPC1P12
DISCRETE I/O MODULES	
Input — 4 pts — 10 to 32 V ac/dc	MPC1M10
Input — 4 pts — 90 to 140 V ac/dc	MPC1M11
Input — 4 pts — 180 to 280 V ac/dc	MPC1M12
Output — 4 pts — 10 to 54 V dc	MPC1M20
Output — 4 pts — 18 to 250 V ac	MPC1M21
Output — 4 pts — Relay (2 "Form C")	MPC1M23
ANALOG I/O MODULES	
Programmable Analog Input Module — 2 Channels	MPC1M13
Programmable Analog Output Module — 2 Channels	MPC1M22

2

T ECHNICAL
I NFORMATION
P UBLICATION
F200
(MPC1)

CUTLER-HAMMER PROGRAMMABLE CONTROLLERS
— 7/1/86
MPC1 Modular Programmable Logic Controllers

Product Description	Catalog Number
MISCELLANEOUS	
Printer w/Cable	MPC1P11
Cable for Printer (Cable Only)	MPC1A29
Register Access Module w/Cable	MPC1P20
Expansion Cable w/connectors for local Expander	MPC1A10
Expansion Connector for remote Expanders	MPC1A11
Power Supply for Programmer	MPC1A12
Replacement Battery (Do not substitute)	MPC1A13
Filler Module	MPC1A14
Fuse for MPC1M20/M21	MPC1A15
Terminal Block Replacement w/Labels	MPC1A16
Spare Terminal Block Labels	MPC1A17
Spare I/O Labels	MPC1A18
Spare Expander Labels	MPC1A19
EEPROM Burner Cartridge (EEPROM not included)	MPC1A20
Replacement EEPROM	MPC1A21
Program Storage Cartridge	MPC1A22
Spare Bus Terminator	MPC1A23
Spare Keying Tabs	MPC1A24
Fuse for MPC1C10/C12/C20/C22 (Do not substitute)	MPC1A25
Fuse for MPC1C30	MPC1A26
MOV Surge Suppressor	MPC1A27
Demonstrator Carrying Case	MPC1A28
Loop Resistor Module (For Analog I/O Module)	MPC1A30
Additional Manuals — NI-117 and NI-164 (Analog)	MPC1D10

SPECIFICATIONS

Component	Dimensions		
	Length	Width	Height
Programmable Controller	14.02" (356.11 mm)	9.61" (244.09 mm)	5.81" (147.57 mm)
I/O Expander Chassis	14.02" (356.11 mm)	7.87" (347.47 mm)	5.81" (147.57 mm)
Programmer	8.59" (218.19 mm)	5.52" (140.21 mm)	1.90" (48.26 mm)
Program Storage Cartridge	2.8" (119.38 mm)	2.02" (51.31 mm)	0.84" (21.34 mm)
EEPROM Cartridge	3.69" (79.89 mm)	2.02" (51.31 mm)	1.08" (27.43 mm)

COMPONENT	WEIGHT
Programmable Controller with 32 I/O	12 lb 14 oz (5.84 kg)
I/O Expander Chassis with 32 I/O	10 lb 4 oz (4.65 kg)
Input Module	7.03 oz (199.30 g)
Output Module	7.50 oz (212.62 g)
Program Storage Cartridge	1.5 oz (42.52 g)
EEPROM Cartridge	1.5 oz (42.52 g)

Power Requirements .... 115 Vac, 50/60 Hz ± 20%, 50 Watts
230 Vac, 50/60 Hz ± 20%, 50 Watts

Environmental Requirements
CPU Ambient .... 0° — 60°C
Programmer Ambient .... 0° — 40°C
Relative Humidity .... 5 — 95%

APPROXIMATE DIMENSIONS
Dual Dimensions — Inches (mm)

Do not use for construction.

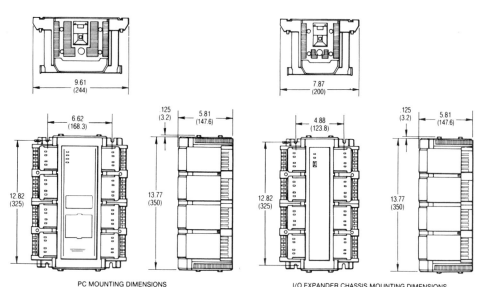

PC MOUNTING DIMENSIONS

I/O EXPANDER CHASSIS MOUNTING DIMENSIONS

See note on installation and use of product at bottom of page 1.

E·T·N

2

Printed in U.S.A.
GG

LARGE PCS

Square D–Your Automation Foundation ™

TECHNICAL OVERVIEW

TO-107-01
JULY, 1986

SY/MAX ® **CLASS 8020 TYPE SCP-721, 722, 723, 724**
MODEL 700 LOGIC CONTROL UNIT
CLASS 8020 TYPE SMM-710, 720
MODEL 700 BUBBLE MEMORY MODULE
MEMORY CONTROL UNIT

The SY/MAX Model 700 is a very powerful programmable controller designed for medium to large applications. The Logic Control Unit (LCU) is enclosed in a sheet metal steel case, and contains processor logic for storing the control program. The Memory Control Unit (MCU) is also enclosed in a sheet metal case and provides non-volatile backup utilizing bubble memory. The Model 700 LCU and MCU are designed to plug into processor and adjacent register slot of any SY/MAX register rack assembly.

The Logic Control Unit function is to process the user ladder diagram program. This involves obtaining and changing I/O status, solving the logical statement and ladder rungs and processing integer and floating point math operations. Contained in the LCU is a random access volatile memory (8K, 16K, 32K & 64K) where the user ladder control program is stored. The LCU contains five diagnostic indicating LEDs and a block of diagnostic registers for system troubleshooting ease. To maintain system security the processor has a three position (RUN, HALT, or DISABLE OUTPUTS) key operated mode selector switch.

The function of the Memory Control Unit is to manage the four RS-422 compatible communication ports and to store the control program when power to the system is shut off. Note: Each port's communication rate (BAUD) is individually programmable to 19.2K max. The magnetic bubble memory (32K & 64K) provides a non-volatile method of storing not only the ladder control program and system definitions during an off power condition but program labels and comments as well. Three diagnostic LEDs indicate current MCU status.

The SY/MAX Model 700 has the ability to address up to 8000 data registers that reside within SY/MAX Local Interface (LI) Modules. Each register can be assigned as 16 digital I/O or for storing data values from timers, counters, or intelligent I/O modules such as Analog and BCD I/O, High Speed Counter Input, Stepper Motor and Speech Output Modules.

The LI allows for serial communication between the I/O modules and CPU at distances up to 15,000 ft. (4572m). The local interface module registers can be used for external digital and intelligent I/O in addition to internal relays and data storage registers. Each local interface module has the capacity for 2048 digital I/O. Although the Model 700 can support a maximum of seven local interface modules

(14,000 I/O) the actual number of I/O a particular processor can control is dependent on memory size and application requirements. The ability to communicate to multiple LI modules results in improved system throughput times.

The SY/MAX Model 700 has all the functions of the Model 500 processor (ASCII output, peer-to-peer or network communications, asychronous shift registers, hardware and password security, etc.) with many additional features such as: 32 bit floating point math, statistical functions, full set of PID closed loop control algorithms (direct acting, reverse acting, lead/lag and manual) and a full set of trigonometric functions, etc. Refer to the Instruction Set table.

Multi-level hardware/software security features can be used to prevent the alteration of data and control programs, prevent I/O forcing, restrict network access of information, inhibit program viewing and provide password security.

SPECIFICATIONS

Model 700 Processor

Rated current draw of LCU and MCU (combined) on SY/MAX power supply is 7300 mA at 5 VDC. MCU draws an additional 800 mA at 12 VDC.

Battery
(active)................70 mA
(standby)..............100 uA
Environmental:
Ambient Temp
(Operational)..........0° to 60°C
Ambient Humidity
(non-condensing)........0% to 95% RH
Storage Temp...........-40° to 80°C
Vibration..............10-15 Hz at 0.015 in. (max) displacement
Dimensions: without key (W x H x D)
LCU...................1.5 x 12.8 x 9.6 in.
MCU...................1.5 x 12.8 x 9.6 in.
Weight (approximate):
LCU...................4.9 lbs.
MCU...................4.5 lbs.
Scan Speed: (approximate, depends upon content)
1.7 ms/K (Relay Logic Only)
Noise Testing.......Designed to meet NEMA ICS2-230 (Showering Arc) and ANSI C37.90a (Surge Withstand Capability)

Capabilities:
Logic................. Microprocessor-based
LCU Memory Size......SCP-721..........8K RAM
SCP-722........16K RAM
SCP-723........32K RAM
SCP-724........64K RAM
MCU Memory Size......SMM-710.............32K Non-Volatile Bubble
SMM-720.............64K Non-Volatile Bubble
Memory Utilization.....1 word per contact or coil
Memory Overhead......47 words
Diagnostics Lights (Status Indicators):
LCU..................RUN, HALT, MEMORY, FORCE, I/O
MCU..................12 VDC, MEMORY, WRITE PROTECT
Total Addressable
Registers..............8000
External I/O...........Over 14,000. Any mix in groups of 4 or 8 I/O points per module.
Internal Relay
Equivalents...........Over 14,000
Data Storage Registers.. Flexible to 8000, 16-bit data registers or 4000, 32-bit floating point registers.

INSTRUCTION SET

INSTRUCTION	DISPLAY	DESCRIPTION	REMARKS
Contacts	—| |—	Represents on/off status of inputs, outputs, internal relays, and register bits.	Input devices can be monitored in HALT.
Coil	—( )—	Represents an output connected to the processor.	Upon a transition to RUN, all coils are de-energized on the first scan. During the second scan, all coils are updated depending on the solution of the logic in that rung.
Internal Relays	R- —( )—	Represents control relay functions	Operation is much the same as an output but does not directly control output devices.
Latch/Unlatch Relays	—(L)— —(U)—	Duplicates the action of a mechanically held latching relay. This is accomplished through the use of two internal relay coils with the same address. Once a latch coil is energized, it remains on until the unlatch coil is energized, even if power is removed.	Latch and unlatch rungs must be programmed sequentially.
Transitional Coil	—(T)—	An internal or external coil which will only energize on a transition of the logic from open to closed. The output will remain on for a single processor scan. The logic must turn the coil off and transition to on again to re-energize the output coil.	Requires two consecutive I/O addresses, one to record the transition and one to remember the last state. A transitional coil must be assigned an ODD address number. The processor will then use the next I/O address for recording the last state of this same coil.
Master Control Relay	MCR —( )—	MCR ON allows logic to function. MCR OFF disables outputs. MCR controls all succeeding rungs until a new MCR coil is programmed or the end of program is reached.	

TO-107-01

MODEL 700 INSTRUCTION SET (continued)

INSTRUCTION	DISPLAY	DESCRIPTION	REMARKS
Timer	TMR	Timer register can store a four digit decimal value with a range from 0 to 9999. Selectable time bases are .01 sec., .1 sec., .1 min.	Programming techniques allow for programming of ON or OFF delay and interval timers. Replaces timing relays and motor driven timers.
Counter	CTR	Counter register can store a four digit decimal value with a range from 0 to 9999. UP, DOWN, and UP/DOWN counters are standard.	It is possible to cascade counters to obtain a count greater than four digits.
Synchronous Shift Register	SHFT	Used to shift individual bits of a storage register in either a forward or reverse direction. Possible to shift 1, 8, or 16 bit channels at a time.	Used in applications requiring sequencing and replaces drum switches.
Asynchronous Shift Register (First In, First Out)	FIFO	Data is loaded into the first zone and exits out the last zone using IN and OUT commands respectively. Possible to shift 8 or 16 bit channels at a time.	Used to record the order of events as they occur, such as in conveyor and alarm annunciation applications.
Data Comparison	IF $(=, \neq, \geq, <)$	Compare functions between storage registers or between an integer or floating point storage register and a constant. When an IF comparison is true, the IF box acts like a short.	It is possible to program 3 comparisons on the line. Integer mode is capable of comparing a 5 digit number from -32,768 to 32,767. In floating point mode a range of -3.4×10^{38} to 3.4×10^{38} and values as small as $\pm 1.2 \times 10^{-38}$ are allowed. Register compares can support negative values. Math functions can be used in IF statements.
Data Transfer	LET	Transfer data from one storage register to another. Also performs integer or floating point or combined mode math.	Ranges are same as Data Compare (above). Occurs once per scan.
Transitional LET, IF	T LET T IF	Transition sensitive; restricts the operation of the LET or IF instruction to one operation for each open to closed transition of the input condition.	
Indirect Register Read	RDSTAT RDDATA FNDBIT	Indirect register read of status field. Indirect register read of data field. Locate and clear the first bit (lowest bit number) set in the indirect register being read.	
Binary to BCD	BCD	Conversion of binary data in a storage register to the binary coded decimal (BCD) equivalent of that data, within a LET or IF instruction.	Allows the processor to drive a BCD-type LED display.
BCD to Binary	BIN	Conversion of binary coded data (BCD) in a storage register to the binary equivalent of that data. Used in a LET or IF instruction.	Used in applications to convert the on/off status from a thumbwheel switch to a binary number which the processor can operate on.
Standard Math Operations (Integer and Floating Point math) Addition Subtraction Multiplication Division Square Root Absolute Value Sine Cosine LOG10 Ln $_e$ $(Y)^x$	LET, IF + — x ÷ SQRT ABS SIN COS LOG LN Y**X	Math operations are performed left to right. Programmed in IF or LET instructions. Value range for integer math operations: -32,768 to +32,767 Value range for floating point math operations: -3.4×10^{38} to 3.4×10^{38}, not smaller than $\pm 1.2 \times 10^{-38}$	Intermediate results to $\pm 2,147,483,647$ are valid.

MODEL 700 INSTRUCTION SET (continued)

INSTRUCTION	DISPLAY	DESCRIPTION	REMARKS
Advanced Math and Trig Operations	INVERT	Divide 1 by the current result (1/X).	
	NEGATE	Change sign of current result.	
	ROUND	Round floating point number to integer.	
	XSQRD	Square current result.	
	HYPOT	Calculate two-dimensional hypotenuse.	
	HYPXYZ	Calculate three-dimensional hypotenuse.	
	TANGNT	Calculate tangent.	
	COSEC	Calculate cosecant.	
	SECANT	Calculate secant.	
	COTAN	Calculate cotangent.	
	ARCSIN	Calculate arc sine.	
	ARCCOS	Calculate arc cosine.	
	ARCTAN	Calculate arc tangent.	
	ARCCSC	Calculate arc cosecant.	
	ARCSEC	Calculate arc secant.	
	ARCCOT	Calculate arc cotangent.	
	SETDEG	Perform subsequent calculations in degree mode.	
	SETRAD	Perform subsequent calculations in radian mode.	
	DEGRAD	Convert degrees to radians.	
	RADDEG	Convert radians to degrees.	
		The following constants are stored to 80-bit precision:	
	PI	Multiply result by pi	
	L2T	Multiply result by log base 2 of 10	
	L2E	Multiply result by log base 2 of natural log e	
	LG2	Multiply result by log base 10 of 2	
	LN2	Multiply result by log base e of 2	
	EULER	Multiply result by Euler's constant	
	E	Multiply result by natural log e	
	1DEG	Multiply result by 1 degree in radians	
	1RAD	Multiply result by 1 radian in degrees	
	ETOPI	Multiply result by e^{π}	
	PITOE	Multiply result by π^{e}	
	ETOE	Multiply result by e^{e}	
	ETOEULR	Multiply result by e^{Euler}	
	SQRTE	Multiply result by $\sqrt{e}$	
	SQRTPI	Multiply result by $\sqrt{\pi}$	
Statistical Operations	ADDSTA	Store current result in statistical block of registers consisting of the summation of previous values, summation of the squares of previous values, and the accumulated count of previous values.	
	VARNCE	Calculate statistical variance on the accumulated totals in the statistical block.	
PID Computations	PIDR	Calculate reverse-acting PID loop.	
	PIDD	Calculate direct-acting PID loop.	
	PIDM	Calculate PID loop in manual.	
	LEDLAG	Calculate lead/lag function for PID loop.	
Accumulator Manipulations	YTOALT	Same as $(Y)^{x}$ except X in this case is taken from the alternate accumulator instead of the second argument.	These manipulations allow the user to preserve 80-bit precision when performing calculations which extend beyond a single rung.
	SWAP 16	Exchange low and high bytes of the current intermediate result.	
	SWAP 32	Exchange low and high words of the current intermediate result.	
	ACCALT	Exchange current accumulator with the alternate accumulator.	
	ADDALT	Add current intermediate result ot the alternate accumulator.	
	SUBALT	Subtract current intermediate result from the alternate accumulator.	
	DUPALT	Copy alternate accumulator into current accumulator.	
	DUPACC	Copy current accumulator (intermediate result) into alternate accumulator.	
	HYPALT	Same as HYPOT except the alternate accumulator is used instead of the second argument.	
	ATANYX	Similar to ARCTAN except that the function is performed on the ratio Y/X.	

MODEL 700 INSTRUCTION SET (continued)

INSTRUCTION	DISPLAY	DESCRIPTION	REMARKS
Identify Number Type	NUMTYP	Identifies mathematical classification of current intermediate result.	
Forcing	CRT INITIATED	Forcing overrides the actual input status or output as controlled by the ladder diagram.	
Communications to other processors or peripherals	T READ T WRTE T ALRM T PRNT	Transfer storage register and I/O status between devices which may be located up to 10,000 ft. apart. All are transition sensitive.	Requires no additional interface hardware.
Read	T READ	Transfers storage register data from remote device to local processor.	
Write	T WRTE	Transfers storage register data from a local processor to a remote device.	
Alarm	T ALRM	Transfers a numerical value from a local processor to a register in a remote device. The number can range from -32,768 to 32,767.	
Print	T PRNT	Transfers ASCII coded messages from the processor to any device that communicates in ASCII.	Simplified means of producing alarm messages, production reports, etc. ASCII data format is selectable.
AND	Λ	A logical function used within an IF or LET command which has the property such that if X and Y are two logic variables, then the function "X AND Y" is defined by the following table: X Y X AND Y 0 0 0 0 1 0 1 0 0 1 1 1	Integer mode only.
OR	$\vee$	A logical function used within an IF or LET command which has the property such that if P and R are logic quantities, then the logical result of "P OR R" assumes values as defined in the following table: P R P OR R 0 0 0 0 1 1 1 0 1 1 1 1	Integer mode only.
Exclusive OR	$\oplus$	A logical function used within an IF or LET command having the property such that if S and T are logic quantities, then the logical result of "S XOR T" is defined by the following table: S T S XOR T 0 0 0 0 1 1 1 0 1 1 1 0	Integer mode only.
Random Number Generator	RANDOM	Generates random number.	
GOTO	GOTO	Instruction to cause the processor to skip over a section of the program to a designated position in the program. Position is determined by a MARK rung placed in memory.	Shortens processor scan time and immediately jumps to a designated position.

TO-107-01

MODEL 700 INSTRUCTION SET (continued)

INSTRUCTION	DISPLAY	DESCRIPTION	REMARKS
Subroutines	GOSUB MARK ST SUB RTN	Group of ladder diagram rungs that can be executed from the user's ladder diagram only by enabling a GO SUB instruction which is programmed in the main program. Subroutines consist of a start of a subroutine (MARK), one or more ladder rungs, and a return (RTN). All subroutines must be programmed after the ST SUB instruction which denotes the beginning of the subroutine area. Nesting (subroutines within subroutines) are allowed up to 8 levels deep.	Applications include variable sequence machines, recipe programs, repeated programs.
Timed Interrupt, Communications Inhibit, Scan Time Limit	Reg. 8165	A special safeguard rung using register 8165 allows the following features: 1. Communications Inhibit 2. Program Viewing Inhibit 3. Safeguard Registers 4. Timed Interrupt-Scan Control 5. Scan Time Limit	Protect Programs used in conjunction with 8176 ⊢( )⊣ - 16 Scan Control
Matrix	M LET M IF M	A matrix is a group of storage registers handled as a unit. Formulates multi-register data transfers and math functions. Also allows multi-register compare statements.	Application includes machine diagnostics, menu selection, etc.
Array	A LET A IF A	An array is a group of 32 bit floating point data storage registers operated on as a unit. Formulates multi-register data transfers, compares and math functions.	
Immediate Communications Update	Reg. 8176 —( )— -11, 12, 13 and 14	Processes incoming data values from COMM port channels which were received during processor scanning.	

ORDERING INFORMATION

CLASS	TYPE	DESCRIPTION
8020	SCP-721	Model 700 Processor Module with 8K RAM Memory
8020	SCP-722	Model 700 Processor Module with 16K RAM Mmeory
8020	SCP-723	Model 700 Processor Module with 32K RAM Memory
8020	SCP-724	Model 700 Processor Module with 64K RAM Memory
8020	SMM-710	Model 700 Memory Module with 32K Magnetic Bubble
8020	SMM-720	Model 700 Memory Module with 64K Magnetic Bubble

For additional information refer to Instruction Bulletin 30598-107-XX.

CUTLER-HAMMER PROGRAMMABLE CONTROLLERS

7/1/86
New TIP

D500 Series Programmable Logic Controllers

T ECHNICAL
I NFORMATION
P UBLICATION
F200
(D500)

1

CONTROLLER

The two D500 Processors — CPU25 and CPU50 — offer RAM or PROM memory, ladder logic programming, fast scan time and a large instruction set. Timers, counters, bidirectional shift registers, master control and jump functions are available. In addition, there is four function math with compare, square root, trigonometric functions, averaging and absolute values. Data transfer, data conversions and enhanced self-diagnostics are included as standard functions. The CPU25 is capable of 256 I/O points. The CPU50 is capable of 512 I/O points.

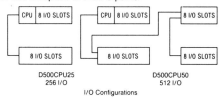

D500CPU25
256 I/O

D500CPU50
512 I/O

I/O Configurations

NETWORKING

The EasyNet data highway system allows multiple D500 units to be linked together. Peer-to-peer communication between D500 controllers and between controllers and remote I/O stations allow for system flexibility. As many as 16 units, controllers or remote I/O stations, can exchange data on one EasyNet system.

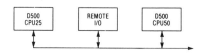

- 16 Stations Maximum
- 1 km [0.6 mi] Total Distance
- 187.5 KBPS Transmission Speed

DESIGN CHARACTERISTICS

The D500 is a full function Programmable Logic Controller offering all of the capabilities of larger frame PLC's in a compact, economical, space saving design. The D500 features:

- Choice of two processor
- Discrete and Analog I/O
- RTD and High-Speed Inputs
- Networking and Computer Interface capability
- Easy-to-use software
- Rugged construction for industrial service

COMPUTER INTERFACE

The computer Interface Capability of the D500 allows an IBM or compatible PC to be used as a program panel. The software, along with the computer interface modules, allow programming, data acquisition, trouble shooting, including power-flow and extensive documentation.

The Computer Interface Capability also allows for control of, supervision of, and data acquisition from multiple PLC's and EasyNet systems.

I/O MODULES

A complete family of I/O modules is available:

- Discrete modules with 16 and 32 points
- Analog input and output modules
- RTD modules for temperature control
- High-speed counting to 50 Khz

COMPONENTS

Description	Catalog Number
SYSTEM COMPONENTS	
Central Processing Unit — 256 I/O, 4k, incl. 8 Slot I/O Rack	D500CPU25
Central Processing Unit — 512 I/O, 8k, incl. 8 Slot I/O Rack	CPU50
8 Slot I/O Expansion Rack	RAC8
8 Slot I/O Expansion Rack with Power Supply	RPS8
Data Access Panel	DAP10
Graphic Program Panel	GPP40
INPUT/OUTPUT MODULES	
Discrete Input Modules — 16 Point, 24 V dc	D500DIM1624D
32 Point, 24 V dc	DIM3224D
16 Point, 115 V ac	DIM1615A
16 Point, 230 V ac	DIM1623A
Discrete Output Modules — 16 Point, 24 V dc	DOM1624D
32 Point, 24 V dc	DOM3224D
16 Point, 115 V ac	DOM1615A
16 Point, Relay	DOM1600R
16 Point, 230 V ac	DOM1623A
Analog Input Modules — ±5 V dc	D500AIM205
±10 V dc	AIM210
±20 ma	AIM220
Analog Output Modules — ±5 V dc	AOM205
±10 V dc	AOM210
±20 ma	AOM220
RTD Input Platinum 100 ohms	D500RTD10
RTD Input Nickel 500 ohms	RTD50
IBM Compatible Software for Programming, Data Acquisition, Documentation and Trouble Shooting	CCS60
IBM Computer Compatible Software Demonstration Disk	CCS10D

2

T ECHNICAL
I NFORMATION
P UBLICATION
F200
(D500)

CUTLER-HAMMER PROGRAMMABLE CONTROLLERS

D500 Series Programmable Logic Controllers

7/1/86

Description	Catalog Number
ACCESSORIES	
Pulse Count Module — 50 khz	D500PCM50
EasyNet Communication Module............	ECM16
EasyNet Remote I/O Station with Communications and 8 Slot I/O Rack	ERS16

APPROXIMATE DIMENSIONS
Dual Dimensions — Inches (mm)

Do not use for construction.

Power Requirements
Processor .... 115 V ac, 50/60 Hz, 50 VA
.... 230 V ac, 50/60 Hz, 50 VA
Programmer .... 115 V ac, 50/60 Hz, 20 VA

Environmental Requirements
Processor .... 0°–55°C operating
Programmer .... 0°–40°C operating
Relative Humidity .... 20%–90% noncondensing

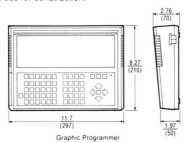

Graphic Programmer

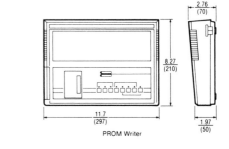

PROM Writer

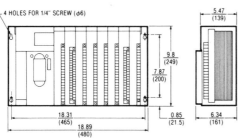

Basic Controller — D500❶
& EasyNet Remote I/O Station

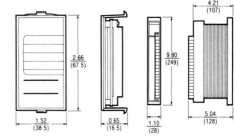

PROM

I/O Module

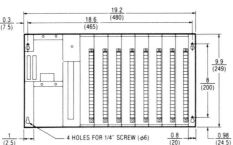

Expansion I/O Rack — With 5 V dc Power Supply❶

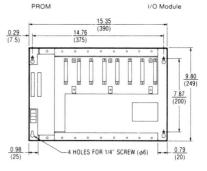

Expansion I/O Rack Without Power Supply

❶ *Shown with full complement of I/O Modules installed*

See note on installation and use of product at bottom of page 1.

PERIPHERALS AND ACCESSORIES

Square D–Your Automation Foundation ™

TECHNICAL OVERVIEW

TO-257-01
SEPTEMBER, 1986

SY/NET˚ CLASS 8030 TYPE CRM-510 NETWORK INTERFACE MODULE

The SY/NET® Local Area Network is a high speed industrial communication system which allows up to 200 devices to intercommunicate over a single twinaxial cable. The network utilizes a unique masterless time token passing system with variable length message packets. This method allows high data throughput rates and predictable access times. The SY/NET network is controlled by SY/NET Class 8030 Type CRM-510 Network Interface Modules (NIM's). Up to 100 NIM's can be connected to a single network and each NIM can support two devices. The NIM's also allow interconnecting multiple networks together.

The NIM resides in the register slot of a SY/MAX® I/O rack assembly and provides a NET port, located beneath the module, and two COMM (communication) ports on the front of the module. Through these ports the NIM supports two types of communication: network communication (NIM-to-NIM), and COMM port communication (PC- or PC, CRT-to-PC, etc.) The NIM coordinates communication between the COMM ports and the NET port. Network connectable devices include SY/MAX Model 100, 300, 500 and 700 Processors, the SY/MAX Process Control Module, D-LOG Data Controller Modules, CRT's, computers, printers, modems, bar code readers, and other ASCII devices.

Network communication consists of a synchronous data stream from one NIM to another. This data structure is in the form of a protocol known as HDLC (High Level Data Link Control). HDLC provides data transfer, message acknowledgements, and error recovery. All NET ports communicate to each other using HDLC. COMM port communication consists of an asynchronous data stream from the NIM to a processor, CRT or other device. The data structure is similar to ANSI 3.28 protocol. This data can refer to ladder rungs, register data, I/O status, and other information.

The SY/NET network allows register data, I/O and processor status to be transferred between individual processors using simple communication function commands (READ, WRITE, and ALARM). Special broadcast functions allow sending messages to all or a select group of devices over the network. SY/NET network capabilities also include using a SY/MAX CRT Programmer or personal or industrial computer, with a SY/MATE™ Software Support package, over the network. This enables all processor programming (up loading, down loading, altering), monitoring, data entry, I/O forcing, and diagnostic functions to be performed from a central programmer or several programmers.

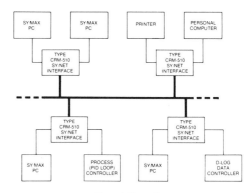

Example Network

NEW LISTING
Printed in U.S.A.

AJR

In order to direct communication from one device to another over the network, each device on the network is assigned a network device number by setting the thumbwheel switches on the face of the module. Once defined, two or more network device numbers can be added to communication rungs creating a from-to path (or ROUTE) which the message will take over the network.

Before a message is sent, the NIM will hold the message and wait its turn to communicate, thereby preventing message collisions. Network access is controlled by a time sharing method which is determined by the network device number assigned. The time sharing method allows every device a chance to "talk" on the network, which prevents any NIM from dominating the network.

The NIM has eight diagnostic LED indicators - network activity, received data Port 0, transmitted data Port 1, network receive error, network transmission error, and power supplied. Baud rate selections for the NET and COMM ports are set by DIP switches accessible on the side of the module. Network transmission speed is dependent on the cable length, as indicated in the following information.

Devices which can be connected to a SY/NET Local Area Network include:

- SY/MAX Model 100, 300, 500 and 700 Programmable Controllers
- SY/MAX Process Control Module
- SY/MAX CRT Programmer
- SY/MAX D-LOG Data Controller
- SY/MAX Speech Module
- IBM (or compatible) Personal and Industrial Computers
- Digital Equipment Corporation PDP and VAX Computers
- Various Color Graphic Monitors
- ASCII Peripherals (Printers, CRT Terminals, etc.)
- PRŌVOX PCIU Distributed Process Control System (Requires Class 8030 Type CRM-530 Network Interface Module. Refer to Technical Overview TO-259-01.)
- Honeywell TDC Distributed Process Control System (Requires Class 8030 Type CRM-570 Network Interface Module. Refer to Technical Overview TO-151-01.)
- SY/NET Remote Area Communication Network. (Requires Class 8030 Type CRM-560 Remote Interface Module. Refer to Technical Overview TO-150-01.)
- MODBUS (Requires Class 8030 Type CRM-530 which implements 3 MODBUS commands or Type CRM-570 which implements 7 MODBUS commands. Refer to Technical Overviews TO-259-01 and TO-251-01 respectively.)

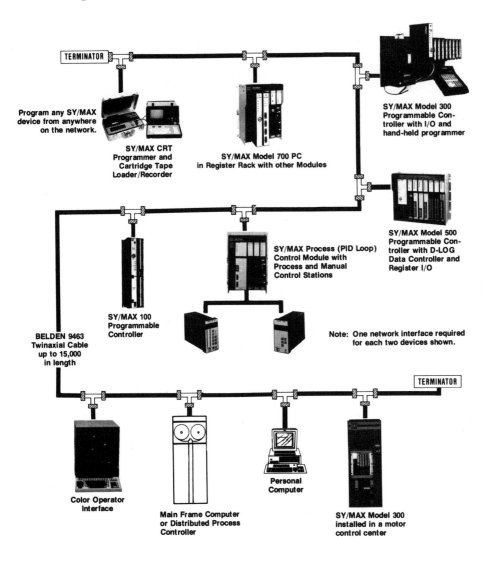

TERMINATOR

Program any SY/MAX
device from anywhere
on the network.

**SY/MAX CRT
Programmer and
Cartridge Tape
Loader/Recorder**

**SY/MAX Model 700 PC
in Register Rack with other Modules**

**SY/MAX Model 300
Programmable Con-
troller with I/O and
hand-held programmer**

**SY/MAX Model 500
Programmable Con-
troller with D-LOG
Data Controller and
Register I/O**

**SY/MAX Process (PID Loop)
Control Module with
Process and Manual
Control Stations**

**SY/MAX 100
Programmable
Controller**

**BELDEN 9463
Twinaxial Cable
up to 15,000
in length**

Note: One network interface required
for each two devices shown.

TERMINATOR

**Color Operator
Interface**

**Main Frame Computer
or Distributed Process
Controller**

**Personal
Computer**

**SY/MAX Model 300
installed in a motor
control center**

TO-257-01

SPECIFICATIONS

Network Specifications

Cable .... Baseband twinaxial cable, Belden 9463 or equivalent

Connectors .... Dual concentric twist lock. Tees, cable ends, adapters, etc. available from Square D

Terminators .... Required on each end of the network cable

Clock frequency .... 1 megahertz

Cable length .... Communication baud rate dependent (cable length vs BAUD rate)

Transmission type .... Manchester encoded HDLC

Message format .... Packet type similar to ANSI 3.28

Collision avoidance .... Token passing, based on time windows

Error checking .... CRC (cyclic redundancy code) on all network transmissions

Message routing .... Source/destination message routes are placed in the command for each device wishing to communicate

Number of devices .... 200 devices per network maximum

Expansion .... Net-to-Net linking - standard, no limit

Interface Module Specifications

Power .... 5 VDC from SY/MAX power supply

NET port .... Single connector. Baud rate switch selectable to 500K Baud.

COMM ports .... Two independent RS-422 receptacles per module. Baud rates switch selectable to 9600 Baud.

COMM port modes .... Four - SY/MAX devices, 8881 Programmable Controller, Net-to-Net, and peripheral devices.

Addressing .... Module address selected by tool operated thumbwheel switches

Indicating LEDs .... Eight

Dimensions (WxHxD) ...... 1.5 x 12.8 x 6.6 inches 3.81 x 32.5 x 16.76 cm

Ambient Temperature Rating .... 0 to 60°C (operational)

Humidity Rating .... 0 to 95% relative humidity (non-condensing)

ORDERING INFORMATION

CLASS	TYPE	DESCRIPTION
8030	CRM-510	SY/NET Local Area Network Interface Module

For additional information refer to Instruction Bulletin 30598-257-01.

Square D–Your Automation Foundation™

TECHNICAL OVERVIEW

TO-317-01
SEPTEMBER, 1986

SY/MATE™ CLASS 8010 TYPE SYM-324 SY/MATE 3 SOFTWARE SUPPORT SYSTEM

The SY/MATE 3 Software Support System is a hardware and software based programming package for all SY/MAX® Class 8020 Processors, Class 8030 D-LOG Data Controllers, and Class 8040 PID Loop Controller. SY/MATE 3 can also interface with the SY/MATIC™ Welder Control. Like all SY/MATE systems, SY/MATE 3 uses an IBM® or 100% compatible personal or industrial computer as the program interface.

SY/MATE 3 incorporates all the features of SY/MATE 1 and 2. These include:

1. On-line program development and monitoring for all SY/MAX Class 8020 Processors (Model 100, 300, 500 & 700).

2. The ability to enter the PID loop tuning parameters of the SY/MAX Class 8040 PID Loop Controller.

3. Screen-prompted programming instructions executed through multi-function soft keys.

4. Off-line program development.

5. Copy all or part of off-line memory to processor memory and vice versa (TRANSFER function).

6. Programming and monitoring by I/O LABELS (function names).

7. Advanced I/O diagnostics using the TIME SCAN (I/O histograph) function.

8. All programming, monitoring and disk functions can be performed via the SY/NET® Local Area Network (LAN).

9. Annotated program documentation, with 18 character I/O LABELS and full page rung COMMENTS.

SY/NET LOCAL AREA NETWORK

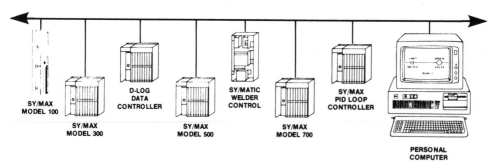

SY/MAX
MODEL 100

SY/MAX
MODEL 300

D-LOG
DATA
CONTROLLER

SY/MAX
MODEL 500

SY/MATIC
WELDER
CONTROL

SY/MAX
MODEL 700

SY/MAX
PID LOOP
CONTROLLER

PERSONAL
COMPUTER

IBM is a registered trademark of International Business Machines Corporation.

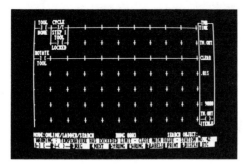

SY/MATE Annotated Documentation

In addition to the powerful functions previously mentioned, SY/MATE 3 has the following features:

Enhanced Alarm Capabilities

SY/MAX Programmable Controllers (Deluxe Model 300, Model 500 and 700 processors) have the ability to send a coded (data) alarm message to any SY/MAX device. SY/MATE 3 expands on this powerful feature by allowing you to transform the coded alarm information into a 70 character English language message with system data of your choosing. This ensures the operator will understand the meaning of the alarm and take the appropriate action.

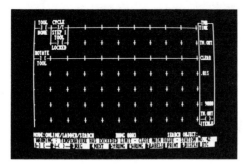

Alarm Screen

The true alarm monitoring power of SY/MATE 3 is exhibited when used with the SY/MATIC High Frequency-DC or AC Welder Control. This device has three levels of alarms, all of which are accommodated by SY/MATE 3 in the method previously mentioned. In addition, two of the three levels allow operator acknowledgement through the program interface keyboard. Since the SY/MATIC welder is SY/NET compatible, a single SY/MATE 3 system provides effective operator interaction with many programmable controllers and welder controls.

Password Security

SY/MATE 3 expands on the four security access levels (MONITOR, DATA ENTER, PROGRAM & OVERRIDE) of SY/MATE 1 and 2 systems. SY/MATE 3 requires an operator to enter a unique password (20 characters max. which you previously defined) before allowing the operator to alter the contents of controller data registers or program system memory. This ensures system security.

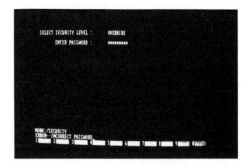

Password Screens

On-Line/Off-Line Programming of the D-LOG Data Controller

The D-LOG Data Controller was designed to have the power of a microcomputer with the resiliency to withstand the factory floor environment. Programmed in Extended BASIC, it communicates with SY/MAX controllers (direct or via SY/NET LAN) to provide an easy and inexpensive means of generating alarm messages, production reports, and graphic displays.

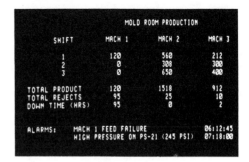

D-LOG Generated Production Report

SY/MATE 3 allows D-LOG programs to be developed on-line (direct cable connection to the D-LOG Data Controller) or in the off-line (computer) memory. Programs stored in off-line memory or on disk can be downloaded into D-LOG memory or vice versa.

File Transfer Between Disk and Cartridge Tape

This feature allows any SY/MAX Processor or D-LOG Data Controller program stored on floppy disk to be transferred to the SY/MAX Class 8010 Type SLR-100 Cartridge Tape Loader/Recorder or vice versa. This provides a more plant-hardened means to backup and transport control programs.

SY/MATE 3 Configuration

The SY/MATE 3 system consists of the 8010 SFW-324 Programming Software (on a single 5-1/4 inch floppy disk) and Interface Board (8010 SFI-324). The interface board is a single circuit board which plugs into a short expansion slot of the IBM Personal or Industrial computer. It provides a nine pin SY/MAX compatible RS-422 differential port. The differential port is electrically isolated from the personal computer and the baud rate is switch selectable. The interface board provides definite advantages over typical serial boards by reducing the risk of computer damage caused by factory induced electrical noise. It also enables the personal computer to be located 10,000 feet (3048m) from the SY/MAX or SY/MATIC device.

SY/MATE 3 is also available as an integrated package which includes the Class 8010 Type SFW-401 Screen-Ware2™Color Graphic Interface. The ScreenWare2 Color Graphic Interface allows the development and operation of real-time animated color graphic displays of manufacturing operations and processes. For more information on ScreenWare2, refer to Technical Overview TO-307-01.

PERSONAL COMPUTER HARDWARE AND SOFTWARE REQUIREMENTS

The personal computer requirements for SY/MATE 3 are:

- IBM PC®, XT®, AT™, Portable, 5331, COMPAQ™ Portable, AT& T Model 6300 or other 100% compatible computer.

- Minimum of one 5-1/4 inch dual sided diskette drive.

- Minimum 384K of memory.

- IBM 1504900 Monochrome Display Adapter or IBM 1504910 Color Graphics Monitor Adapter (or equivalents).

- Monochrome or Color Monitor.

- PC/DOS Version 2.0, 2.1, 3.0, 3.1, or MS/DOS equivalent.

ORDERING INFORMATION

CLASS	TYPE	DESCRIPTION
8010	SYM-324	SY/MATE 3 Software Support System (Includes: SFW-324 software and SFI-324 interface board)
8010	SYM-424	SY/MATE 3 Integrated Programming Package with ScreenWare2 Color Graphic Interface (Includes: SFW-324 software, SFI-324 interface board and SFW-401 ScreenWare2 Color Graphic Interface)

1. An SFI-324 Interface Board will operate with either SY/MATE 1, SY/MATE 2 or SY/MATE 3 Programming Software.

2. Each SFI-32X Interface Board requires a Class 8010 Type CC-100 (10 ft.) or CC-101 (30 ft.) cable. Order separately.

3. ScreenWare2 operates with an IBM PC, XT, AT, 5531, 7531, 7532 or other 100% compatible computers. It requires a color monitor and dual 5-1/4 inch disk drives.

For additional information on SY/MATE 3, refer to Instruction Bulletin 30598-317-XX.

TO-317-01

Square D—Your Automation Foundation™

TECHNICAL OVERVIEW

TO-251-01
NOVEMBER, 1986

SY/MAX® CLASS 8030
TRANSFER INTERFACE SYSTEM

CLASS 8030 TYPE CRM-230
LOCAL TRANSFER INTERFACE MODULE

CLASS 8030 TYPE CRM-232
REMOTE TRANSFER INTERFACE MODULE

The SY/MAX Class 8030 Transfer Interface System enables redundant SY/MAX Model 500 or Model 700 Processors to control a common I/O system. The system will transfer control of the I/O system to a backup processor should the primary shut down for any reason. Designed primarily for use in critical control applications, typical transfer system installations include petroleum refining, petrochemical processing, coal-fired electric utilities, municipal waste water treatment facilities, and other operations which cannot be interrupted by a processor, power or communication malfunction.

The SY/MAX Transfer System does not require fixed primary and backup designations. If a fault condition occurs in the primary processor, control is automatically switched to the backup. The failed primary assumes a failed backup status, and the backup becomes the primary. The fault can then be corrected in the failed backup and activated to operational backup status without interrupting control operations.

A SY/MAX Transfer Interface System consists of two or more Class 8030 Type CRM-230 Local Transfer Interface (LTI) Modules and one or more Class 8030 Type CRM-232 Remote Transfer Interface (RTI) Modules. See figure 1. The LTI modules can designate the primary and backup status of the processors and supervise the actual transfer of control if a fault occurs. The LTI's are also connected to (and control) the I/O devices through the RTI modules.

The transfer system is configured using two Class 8030 Type RRK-100 or RRK-200 rack assemblies and two Model 500 or Model 700 processors. Each processor rack includes at least one LTI module. See figure 2. A cable connection is made between the Register Transfer Channel (RTC) terminal on the front of the LTI modules. The RTC link is used for scan synchronization and data exchange between the two processors. All communication on this channel occurs at 375K baud at a distance up to 25 feet.

The upper section of the wiring terminal is the I/O Channel. A cable run is made from each of the LTI I/O Channels to isolated channels (A and B) on an RTI module located in the first I/O drop. Dual cable runs from the first RTI are then made to an RTI module in each successive drop. Up to eight I/O drops may be made from an LTI pair. Communication between the LTI pair and RTI's is 62.5K baud at a distance of up to 2500 feet. Both LTI modules receive I/O status information simultaneously from the RTI's. The

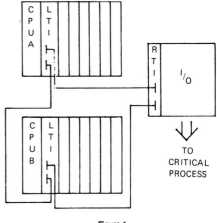

Figure 1
Basic System Configuration

NEW LISTING
Printed in U.S.A.

AJR

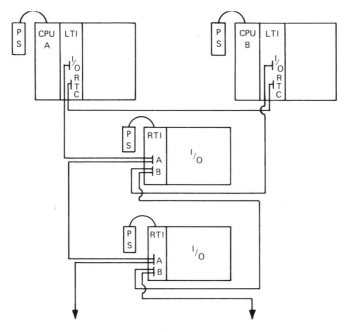

Figure 2
Multiple I/O Drop Configuration

two isolated RTI channels and dual cable runs provide communication and cable redundancy within the transfer system. All cables are shielded dual twisted pair. The module terminal blocks are removable to facilitate module replacement.

With each LTI pair capable of supporting eight I/O drops, a maximum of 1024 I/O can be controlled by the LTI pair. As additional I/O are required, LTI pairs are added, with a maximum of seven pairs possible. The Model 500 processor is capable of addressing 2000 I/O. Redundant Model 700 processors can address a maximum of 7168 I/O when using the transfer system.

Two types of transfers can be performed by the SY/MAX Transfer Interface System. The first is the bumpless transfer which occurs when the backup processor is synchronized with the primary. Processor scans, I/O states and data values are maintained to be the same in both processors. To ensure that critical timer, counter and other data values are identical in both primary and backup processors, 32 registers can be transferred at the end of every processor scan.

If synchronization cannot be maintained between the processor scans (typically due to noise or loss of communication between the LTI modules) a bump transfer condition exists. Although some differences in I/O values may exist at the time of the transfer, a complete shut down is avoided. If a bump transfer cannot be tolerated in a particular application, the backup can be programmed to shut down if synchronization with the primary is lost.

On the upper front of the LTI and RTI modules are seven LED indicators providing precise status and error indication. Located below the LED's on the LTI module is a RESTART push button which can be used to restore lost communication on the I/O or RTC Channels.

Additional redundancy can be incorporated within a transfer system by use of dual power supplies at the processor rack assemblies and at each of the remote racks. See figure 3. Redundant power supplies are connected to a rack through a Class 8030 Type CC-51 Redundant Power Supply Cable. This "Y" cable connects directly to the rack on one end and to two power supplies on the two legs. For additional information on SY/MAX Power Supplies, refer to Technical Overview TO-156-XX.

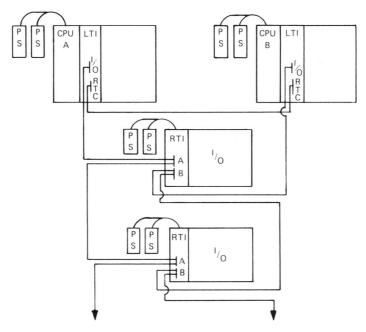

Figure 3
Transfer System With Redundant Power Supplies

SPECIFICATIONS

Electrical Specifications

Rated Current Draw
On SY/Max Power
Supply . . . . CRM-230: 1600 mA
CRM-232: 1000 mA
Registers (CRM-230) . . . . 4096 (255 for external drop addressing, 3841 for internal data storage)
I/O Channel . . . . 1 I/O Channel per LTI
255 Registers maximum
1024 Digital I/O maximum
8 Drops maximum
Drops . . . . 128 Digital I/O maximum
127 Registers per drop maximum
Register Transfer
Channel . . . . 1 Register Transfer Channel per LTI
32 Registers transferrable per scan
Up to 4096 Registers transferrable upon system startup
Communication Method . . . . Continuous full duplex serial differential (RS422A)
Transmission Rate
I/O Channel . . . . 62.5K baud
Transmission Rate
Register Transfer
Channel . . . . 375K baud

Transmission Distance
I/O Channel . . . . 2500 feet (762 meters)
Transmission Distance
Register Transfer
Channel . . . . 25 feet (7.62 meters)

Physical Specifications

Weight (approximate)
CRM-230, CRM-232 . . . . 3.0 lb./1.36 kg.
Dimensions (W x H x D)
CRM-230, CRM-232 . . . . 1.5 x 12.8 x 6.6 in.
(3.81 x 32.5 x 16.76 cm)
Rack Compatibility CRM-230: RRK-100, RRK-200
CRM-232: CRK-210, CRK-310, DRK-210, DRK-300, GRK-110, GRK-210, HRK-100, HRK-150, HRK-200, RRK-100, RRK-200

Environmental Specifications

Ambient Temperature . . . . . . 0 to 60°C
Storage Temperature . . . . -40 to 80°C
Humidity . . . . 0 to 95% non-condensing

ORDERING INFORMATION

CLASS	TYPE	DESCRIPTION
8030	CRM-230	Local Transfer Interface Module
8030	CRM-232	Remote Transfer Interface Module

For additional information refer to Instruction Bulletin 30598-251-XX.

Glossary
Definitions of Commonly Used PC Terms

ALL OF THE NEW TERMS INTRODUCED IN THE TEXT HAVE BEEN COLLECTED HERE. THIS should provide a convenient reference source for those who are just beginning to learn about programmable controllers.

absorptive law—The following law of Boolean algebra:

$$A(A + B) = A.$$

ADC—An Analog-to-Digital Converter. *See* analog interface.

address bus—A bus used to gain access to a particular location in memory.

algorithm—*See* user program.

analog interface—An I/O interface that converts an analog signal to digital or vice versa. It is an analog-to-digital or digital-to-analog converter.

AND—A logic operation that requires that all inputs be true in order for the output to be true.

application memory—The portion of PC memory architecture that contains data memory and user memory.

application program—*See* user program.

architecture—A specification of how various memory systems are organized and used by the PC to perform control functions.

armature—The moving, metallic component of a relay.

ASCII—An acronym for American Standard Code for Information Interchange, a binary code used in computer and PC keyboard applications.

ASCII interface—An interface that allows the transmission of alphanumeric data between the PC and peripherals using the ASCII code.

assembly language—A symbolic programming language that is not often used for user programs.

associative laws—Two laws of Boolean algebra, as follows:

$$(1)\ A + (B + C) = (A + B) + C;\ (2)\ A(BC) = (AB)C.$$

auxiliary power supply—*See* remote power supply.

base—The number of digits in a number system. Base 10 has ten digits (0 through 9), base 2 has two (0 and 1), etc.

baseband—A type of cable that allows the transmission of only one signal at a time.

BASIC—A high-level, English-statement-type programming language for computers and PCs.

basic language—*See* low-level language.

battery back-up system—A battery-based power supply used to protect volatile memories from an interruption in power supply.

baud rate—The number of bits-per-second of binary data that can be received or transmitted during serial communication.

Baudot code—A binary code used primarily in teleprinter machines.

BCD—*See* binary-coded-decimal.

binary—The base-2 number system. It has only two digits: 0 and 1.

binary-coded decimal—A code that represents each digit from 0 to 9 as a four-bit binary number.

binary codes—Systematic methods for presenting or transmitting binary numbers. These include binary-coded decimal (BCD), the gray code, the Baudot code, and ASCII.

bit—A digit in the binary number system (from *bi*nary digi*t*).

block diagram language—A high-level PC programming language that employs functional block symbols in a ladder format.

Boolean algebra—A mathematical shorthand that describes the outcome of logic operations and combinations of logic operations.

Boolean language—A low-level PC programming language based on symbolic representations of Boolean operations (AND, OR, NOT, NAND, and NOR). It is sometimes called *mnemonic* language.

Boolean operations—Logic operations such as AND, OR, NOT, NAND, NOR, and XOR.

broadband—A type of cable that allows the transmission of two or more signals simultaneously on different channels (frequencies). The cable used for cable TV is broadband cable.

broadcast—A form of polling in which the master PC sends a request to transmit data to all slave PCs simultaneously, and then polls the individual slaves for responses.

buffer—A short-term memory for the temporary storage of data.

bus—A line or group of lines used for data transmission or power distribution.

business-systems local area network—A local area network used to connect office equipment, such as word processors, to a central computer. This type of network allows communications between word processors and computer terminals that are scattered about in various office locations.

byte—A group of eight bits.

C—A high-level programming language.

cable—A shielded, very uniform, wire-based communications medium.

central processing unit—A portion of PC hardware that contains a power supply, microprocessor, and memory. The central processing unit interprets and executes programmed instructions.

chip—Slang for integrated circuit.

clock—A timing device, located in the CPU, which provides synchronizing pulses.

CMOS—Acronym for Complementary Metal Oxide Semiconductor, a semiconductor technology often used in RAM memory.

CMP—In programming languages, an instruction that requires that two numbers or words be compared. It usually is followed by = +, >, or <.

coaxial cable—*See* cable.

code—A systematic method for presenting or transmitting information.

coil—The electromagnet portion of a relay. Also, in ladder diagrams, a symbol for an output.

collision detection—A method of controlling network communications that is popular in business systems networks. Collision detection circuitry senses transmissions on the bus and prevents PCs from transmitting when a signal is on the bus.

common bus—A configuration or topology for serial communication between several PCs.

communication standard—*See* protocol.

communications medium—The physical form that a bus takes (e.g., twisted pair conductors, coaxial cable, etc.).

communications network—A communications system linking several PCs. The objective of a communications network is the sharing of data and system status.

commutative laws—Two laws of Boolean algebra, as follows:

$$(1)\ A + B = B + A;\ (2)\ AB = BA.$$

complement—*See* method of complements.

composite logic operations—Logic operations that are formed from combinations of the three basic logic operations, AND, OR, and NOT. Composite logic operations include NAND, NOR, and XOR.

computer-type language—Any high-level, English-statement-type PC programming language. BASIC is an example.

contact ladder diagram—A schematic shorthand notation that represents the current flow status of control relays and associated equipment.

control contact—In ladder diagram programming language, a contact that, when closed, allows the completion of the instructions in the rung.

control network—A communications system linking a PC and peripherals. The objective of a control network is the complete transmission of data and instructions within one scan of the PC.

control program—*See* user program.

control relay—A relay designed for industrial control applications. These are more rugged in design than other relays, and almost always are double-throw relays (i.e., they contain both N.O. and N.C. contacts). The armature of a control relay usually contains a pair of moving contacts to help dissipate transient voltages.

core memory—A nonvolatile RAM consisting of small magnetic donuts.

CORONET—A proprietary local area network from GEC Automation Projects.

counter—A programming symbol used to energize or de-energize an output after a specified count of events.

cps—Abbreviation for Characters Per Second. *See* printing speed.

CPU—*See* central processing unit.

CRT—An abbreviation for cathode ray tube. The CRT is a combination VDT and keyboard that allows system monitoring and program entry. There are two main types of CRTs: dumb (no microprocessor or memory) and intelligent (containing microprocessor and memory).

DAC—Digital-to-Analog Converter. *See* analog interface.

daisy chain—A configuration or topology for serial communication between a PC and several peripherals or between several PCs.

data bus—A bus used to transmit data previously stored in a particular location in memory.

data highway—*See* local area network.

Data Highway II—A proprietary local area network by Allen-Bradley Co.

data memory—The portion of PC memory that holds data used by the microprocessor in fulfilling its control functions (e.g., preset values). *Also see* RAM.

data processing interface—An intelligent interface that performs prescribed calculations on data received from sensors or other devices connected to the interface.

debugging—The process of locating and correcting errors in a user program.

decimal—The base 10 number system. It has ten digits: 0 through 9.

DeMorgan's Laws—Two laws of Boolean algebra, as follows:

$$(1)\ \overline{(A + B + C)} = \overline{A}\,\overline{B}\,\overline{C};\quad (2)\ \overline{(ABC)} = \overline{A} + \overline{B} + \overline{C}.$$

diagnostic—*See* diagnostic program.

diagnostic program—A program included in the microprocessor that detects failure in communications, system operation, etc., and activates an alarm circuit to signal a failure.

dibble-dabble—A method of converting numbers from the decimal to the binary number system.

discrete interface—An I/O interface used with devices that provide or require discrete (i.e., on or off) signals.

distributed control—*See* multiprocessing.

distributed polling—*See* token passing.

distributed processing—*See* multiprocessing.

distributive laws—Two laws of Boolean algebra, as follows:

$$(1)\ A(B + C) = AB + AC;\ (2)\ A + BC = (A + B)(A + C).$$

documentation—The process of preparing a record of the user program, hardware connections, etc., for future reference.

drop cable—A relatively short length of cable that connects a peer PC modem to a bus.

duplex transmission—*See* full-duplex transmission.

EAROM—Acronym for Electrically Alterable Read-Only Memory, a ROM memory that can be erased quickly by application of a relatively low voltage to a pin on the chip.

EasyNet Data Highway System—A proprietary local area network or data highway offered by Eaton/Cutler-Hammer.

EEPROM—Acronym for Electrically Erasable Programmable Read-Only Memory, memory that is erased with an electric charge.

EIA—Abbreviation for the Electronic Industries Association.

electromagnetic interference—Spurious signals and noise generated by electromagnetic equipment (e.g., pumps, motors, etc.).

EMI—*See* electromagnetic interference.

END MCR—A program instruction that with MCR is used to isolate a portion of a ladder diagram.

END ZCL—A programming instruction that is similar to END MCR and is used with the ZCL instruction.

EPROM—*See* erasable programmable read-only memory.

erasable programmable read-only memory—A nonvolatile memory that can be erased and reprogrammed.

EXCLUSIVE-OR—*See* XOR.

executive—A supervisory program stored permanently in the PC that supervises the operation of the PC. This supervisory function includes control, data processing, and communication with I/O modules and peripherals.

executive memory—The portion of PC memory that contains the executive program (PROM).

false—An off or 0 state.

feedback—A signal sent to the PC that gives the response of the process or system to the PC's instructions.

fiberoptics—A communications medium based on the transmission of visible light instead of audio signals.

flow chart—A schematic diagram of a user program.

frequency-shift keying—A technique used by modems to convert binary voltages into two-frequency audio signals and vice versa.

FSK—*See* frequency-shift keying.

full adder—A binary adding circuit composed of two half-adders and one OR gate that is used to add three digits.

full-duplex transmission—A form of data transmission that allows simultaneous, two-way communication between a PC and a peripheral.

gate—*See* logic gate.

GEnet Factory Local Area Network—A proprietary local area network from GE Fanuc Automation.

GET—A programming instruction that requires the microprocessor to obtain data stored in a particular register.

GOSUB—A computer language instruction equivalent to the jump-to-subroutine ladder instruction.

GOTO—A computer language instruction equivalent to the jump ladder instruction.

GRAFCET—A high-level, functional flow chart programming language.

gray code—A binary code that requires a change in only one bit between successive numbers. It is used primarily in mechanical-to-electrical conversions (e.g., in positioning applications).

half-adder—A binary adding circuit composed of one XOR and one AND gate that is used to add two digits.

half-duplex transmission—A form of data transmission that is two-way (i.e., from the PC to the peripheral and vice versa), but with data transmission occurring only in one direction at a time.

handshaking signal—A pair of signals used in communication between peer PCs. The pair consists of a request-to-transmit signal and a clear-to-transmit signal.

hardware—The physical, electrical, electronic, and mechanical devices that form a PC system.

hexadecimal—The base-16 number system. It has 16 digits: 0 through 9 and A through F.

high-level language—A PC programming language based either on block diagrams or computer languages.

IEEE—Abbreviation for the Institute of Electrical and Electronics Engineers.

industrial relay—*See* control relay.

input interface module—A collection of electronic circuits that provide the signal conditioning and isolation required to connect (or interface) a field input device to the PC's microprocessor.

input/output status memory—The portion of PC memory architecture that is reserved for current input/output statuses (RAM).

intelligent alphanumeric display—A programmable device that can display letters, numbers, messages, and warnings.

intelligent interface—*See* intelligent I/O module.

intelligent I/O module—A programmable I/O module or interface, usually remote, that performs only one control function, such as PID control.

I/O—Abbreviation for Input/Output.

I/O interface—*See* input interface module; output interface module.

I/O module—*See* input interface module; output interface module.

inverter—A NOT gate.

ISO—Abbreviation for the International Standards Organization.

isolation transformer—A transformer installed between the ac power line and the PC. It protects the PC from EMI.

JMP—*See* jump.

JSB—*See* jump-to-subroutine.

jump—A programming instruction that, when activated, requires the microprocessor to skip to the designated program rung.

jump-to-subroutine—Similar to jump, this program instruction requires the microprocessor to skip to the designated subroutine.

K—A measure of memory size; 1K of memory provides the capacity for storage of 1024 words.

ladder diagram—A schematic, shorthand notation that represents the power feed (relay ladder) and current flow (contact ladder) statuses of control relays and associated equipment. Also, a programming language employing ladder diagrams.

ladder language—*See* ladder diagram.

latching relay—A retentive relay containing latch and unlatch coils. The contacts remain in the energized position until the unlatch coil is energized.

latch out—In ladder diagram language, a symbol for an output that represents the energizing of a latching relay's latch coil.

LCD—*See* liquid crystal display.

least significant digit (bit)—In a multidigit (multibit) number, the digit (bit) farthest to the right.

LED—*See* light-emitting diode.

light-emitting diode—A semiconductor that emits light when forward-biased.

line printer—*See* printer.

linear voltage differential transformer—*See* LVDT.

liquid-crystal display—A device that forms numbers and characters using reflected light from liquid crystals.

LOAD—A Boolean programming instruction that symbolizes the initiation of a rung with N.O. control contacts.

LOAD NOT—A Boolean programming instruction that symbolizes the initiation of a rung with N.C. control contacts.

local area network—A communications network or system that allows the transmission of data and system statuses between PCs, or between PCs and intelligent devices, at high speeds and over long distances; often called a data highway.

logic circuits—Electronic circuits that perform the three basic logic operations of AND, OR, and NOT.

logic gate—A logic circuit.

LOLa—A proprietary ladder-type programming language from Furnas Electric Co.

loop—A configuration or topology for serial communication between several PCs.

low-level language—A PC programming language based either on ladder diagrams or Boolean (mnemonic) logic.

LVDT—A sensor that provides an output voltage that is proportional to linear displacement.

M—A measure of memory size; 1M of memory provides the capacity for storage of 1,048,576 words (1M = 1K × 1K).

Manufacturing Automation Protocol—*See* MAP.

MAP—Abbreviation for Manufacturing Automation Protocol, a protocol for data highways or local area networks.

master PC—*See* master-slave local area network.

master-slave local area network—A local area network that is controlled by a master PC or computer. Communications between PCs in the network (slaves) are routed through the master PC or computer.

MCR—Abbreviation for Master Control Relay, a program instruction that, with END MCR, is used to isolate a portion of a ladder diagram.

memory—A portion of the PC's CPU that stores data and instructions.

memory burner—A device that *burns* or impresses, a program onto EPROM memory chips.

memory systems—The various types of memory available for use with PCs. These are generally listed as volatile memory and nonvolatile memory.

memory utilization map—A schematic representation of memory architecture.

method of complements—A method used by computers and PCs to perform binary subtraction.

microprocessor—The integrated circuit and associated support circuitry which performs mathematical, logic, data processing, diagnostic, and control functions. The microprocessor controls all the activities of the PC.

miniprogrammer—A hand-held device used to enter programs into the PC's memory or to edit programs.

minuend—A number from which the subtrahend is to be subtracted.

mnemonic language—*See* Boolean language.

Modbus—A proprietary local area network from Gould Incorporated.

modem—Acronym for MOdulator-DEModulator. It serves as an interface between the PC or intelligent device and a communications medium (wire, cable, etc.). It converts two-level binary voltage from the PC to a two-frequency audio signal for transmission, and vice versa.

most significant digit (bit)—In a multidigit (multibit) number, the digit (bit) farthest to the left.

move—A block diagram language instruction that is the equivalent of the ladder instructions GET and PUT.

multidrop—A peer-to-peer version of the common bus topology.

multiprocessing—The use of several microprocessors to perform control tasks. This reduces the time required to implement a program.

NAND—A composite logic operation that is formed from an AND operation followed by a NOT operation.

N.C.—*See* normally closed.

negative logic—A representation in which a true state is represented by zero voltage and a false state is represented by a positive voltage ($+5$ volts dc).

network adapter module—*See* modem.

nibble—A group of four bits.

N.O.—*See* normally open.

node—A device connected to a local area network.

nonvolatile memory—A memory that retains its contents even when the power supply is interrupted.

NOR—A composite logic operation that is formed from an OR operation followed by a NOT operation.

normally closed—A set of relay contacts that do not pass current when the relay is energized. Also, in ladder diagram programming language, an instruction that indicates that a signal is required to open a contact.

normally open—A set of relay contacts that do not pass current when the relay is de-energized. Also, in ladder diagram programming language, an instruction that indicates that a signal is required to close a contact.

NOT—A logic operation that provides a true output for a false input, and a false output for a true input. It is often called an *inverter.*.

NOVRAM—Acronym for NOnVolatile Random-Access Memory. It employs conventional semiconductor RAM and EEPROM, both fabricated on a single integrated circuit.

numerical data interface—An I/O interface capable of handling data in the form of multiple bits (e.g., BCD inputs and outputs).

NumeriExpress Communications Link—A proprietary local area network from Giddings and Lewis Electronics Company.

octal—The base-8 number system. It has eight digits: 0 through 7.

off-line programming—A type of programming, employing intelligent CRTs or programming devices, that allows a program to be written and edited without having to be connected to the PC.

one-shot contacts—A program instruction that closes or opens a contact for one program scan only. A typical application is in unlatching a latched relay.

on-line programming—A type of programming, usually employing dumb CRTs, that involves making program changes while connected to the PC.

OR—A logic operation that yields a true output when any one input is true.

out—A general symbol in ladder diagram programming language that indicates any output (motor, coil, lamp, etc.).

out NOT—In ladder diagram programming language, a symbol that indicates that an output is de-energized when the current path through a rung is completed.

output interface module—A collection of electronic circuits that provide the signal conditioning and isolation required to connect, or interface, the PC's microprocessor to a field output device.

overflow indicator—A warning light or other alarm used to indicate that the capacity of a register has been exceeded.

parallel transmission—A form of data transmission between a PC and its peripherals. In parallel transmission, each bit of data is carried by a separate transmission line, thus allowing rapid, simultaneous transmission of data.

PC—*See* programmable controller.

peer PC—*See* peer-to-peer local area network.

peer-to-peer local area network—A local area network in which each PC (peer) controls its own communications and takes a turn at controlling the network.

peripheral—An external device connected to the PC, such as a CRT, printer, etc.

personal computer—A small, digital electronic device that can store, retrieve, and process data.

photocell—A sensor that produces a voltage that is proportional to the amount of light detected by the sensor.

PID—*See* proportional-integral-derivative control.

PID interface—An intelligent interface used to provide PID control for various processes that require continuous, closed-loop feedback control.

PLC—*See* programmable logic controller.

polling—Sometimes called addressing, the process by which a master PC asks a slave PC for data.

positive logic—A representation in which a true state is represented by a positive voltage ($+5$ volts dc) and a false state is represented by zero voltage.

power supply—The portion of the CPU that provides power to the microprocessor and memory, and often to I/O modules.

PRE—In programming, a symbol for preset. It is used to preset timers and counters.

primary battery—A nonrechargeable battery.

printer—A device that converts electrical signals into printed numbers and characters.

printing speed—The number of characters per second a printer can print.

process control—The continuous, automatic control of processes or operations.

process measurement—The set of techniques used to obtain the values of process variables to be controlled.

processor—*See* microprocessor.

program—A series of instructions that, when executed, allows the PC to control processes and machines.

program loader—A device used to load or reload a program into a PC's memory. Examples include cassette tape recorders and memory burners.

programmable controller—A digital electronic device that meets the three following criteria: (1) It has a programmable memory, in which instructions can be stored; (2) The instructions stored in the memory are used to implement various functions, such as logic, sequencing, timing, counting, and arithmetical; and (3) The various functions are used to control machines or processes.

programmable logic controller—A primitive, first-generation programmable controller.

programmable read-only memory—A nonvolatile memory that cannot be altered once it has been programmed.

PROM—*See* programmable read-only memory.

proportional-integral-derivative control—A closed-loop feedback control technique, often called PID control. It uses a control signal that is proportional to the sum of the error signal, the integral of the error signal, and the derivative of the error signal.

protocol—A specification of physical details (e.g., number of lines), signal levels, and timing for the standardization of data transmission between a PC and peripherals.

PUT—A programming instruction that requires the microprocessor to store an output in a particular register.

query-response—A form of polling in which the master PC asks the slave PC to transmit data and then waits a specified time for a response.

radix—The base of a number system.

RAM—*See* random-access memory.

random-access memory—A volatile, semiconductor memory that is easy to change (i.e., reprogram). It is used for the storage of input data and application programs.

read-only memory—A nonvolatile semiconductor memory. Generally, read-only memory is difficult, if not impossible, to reprogram.

register—A location in memory for the temporary storage of data, instructions, and information.

regulator—An electronic device that maintains power supply output voltage regardless of load.

relay—An electromagnetic switch. It contains an electromagnet (commonly called a coil), which is used to open or close switch contacts.

relay ladder diagram—A schematic, shorthand notation that represents the power feed status of control relays and associated equipment.

relay ladder programming—*See* ladder diagram.

relay-type instructions—In ladder diagrams these are programming instructions based on relay control systems (e.g., N.C., N.O., latch, unlatch, coil, etc.).

remote I/O modules—I/O modules that are located long distances away from the PC location.

remote power supply—A power supply for remote I/O modules.

repeater—A device that receives a signal and automatically retransmits (or repeats) the signal in amplified form.

RET—A programming instruction that requires the microprocessor to return to the main program after completing a subroutine.

ROM—*See* read-only memory.

rung—One line of program instructions in the ladder diagram programming language.

safety margin—The difference between the value of a variable (e.g., temperature) and the value of the variable that is unsafe (e.g., explosive temperature).

scan—A process by which the microprocessor accepts inputs, manipulates data, and updates outputs on a periodic basis.

scan rate—*See* scan time.

scan time—The length of time required to perform a scan. Usually, it is given in milliseconds (msec or ms) per K of memory.

scientific notation—A method of conveniently representing either very large or very small numbers as powers of ten.

scratch-pad memory—The portion of PC memory that is used for the temporary storage of data (RAM).

secondary battery—A rechargeable battery.

serial interface—A special interface module or cable that provides lines and signals for the serial transmission of data between a PC and peripheral.

serial transmission—A form of data transmission between a PC and its peripherals. In serial transmission, all data bits are carried one after another (serially) on a single transmission line.

set point—The desired operating point (e.g., desired temperature, desired pressure, etc.) of a system or operation.

seven-segment display—An output device that displays numbers. It is composed either of seven-segment light-emitting diodes or seven-segment liquid crystal displays.

simplex transmission—A form of data transmission that allows for one-way communication only (e.g., from a PC to a peripheral).

single-pole double-throw—A relay that opens one contact and closes another when energized.

single-pole single-throw—A relay that either opens or closes a single set of contacts when energized.

skip—Either the MCR or ZCL programming instructions.

slave PC—*See* master-slave local area network.

software—Any program or subroutine used by PC hardware.

SPDT—*See* single-pole double-throw.

SPST—*See* single-pole single-throw.

star—A configuration or topology for serial communication between a PC and several peripherals, or between several PCs.

STARNET—A proprietary local area network from GEC Automation Projects.

State Language—A high-level programming language from Adatek.

subroutine—A self-contained program within a larger program. A subroutine may be used several times during the execution of the main program.

subtrahend—A number that is to be subtracted from a minuend.

SUCONET Field Bus—A proprietary local area network from Klockner-Moeller Corp.

SY/NET—A proprietary local area network from Square D Co.

SYSBUS—A proprietary local area network from Omron Electronics, Inc.

system memory—The portion of PC memory architecture that contains executive memory and scratch pad memory.

tee tap—A device used to connect a PC communication cable (drop cable) to the bus.

temperature-activated switch—A switch that changes state (from N.O. to N.C. or vice versa) when a certain temperature change is attained.

thermocouple—A device that produces a dc voltage that is proportional to temperature.

thumbwheel switch—-A 10-position switch that transmits a BCD code corresponding to the number displayed on the switch.

timer—A programming symbol used to energize or de-energize an output after a specified time interval.

TIWAY ONE—A proprietary local area network from Texas Instruments.

token-holding time—The length of time during which one peer PC controls a network.

token passing—A method of passing the control of a local area network from one peer to another.

topology—The configuration or physical arrangement of nodes in a local area network.

transmission media—The physical components that allow communication between PCs or between PCs and intelligent devices. Major components of transmission media are the modem (or network adapter module) and the bus, or communications medium.

triaxial cable—*See* cable.

true—An on or 1 state.

truth table—A table that describes the outputs of a particular logic operation for all possible inputs.

TTL—Acronym for Transistor-Transistor Logic.

twisted-pair conductors—A communications medium composed of two single conductors that are twisted about each other.

unidirectional transmission—*See* simplex transmission.

unlatch out—In ladder diagram language, a symbol for an output that represents the energizing of a latching relay's unlatch coil.

user-friendly—Slang for *easy to use.*

user memory—The portion of PC memory that contains the user program (RAM, EPROM, EAROM).

user program—A program that provides the instructions for a specific control application.

UV-EPROM—An EPROM memory that is erased upon exposure to ultraviolet light.

VDT—*See* video display terminal.

video display terminal—A monitor (CRT screen) that displays program or system status.

volatile memory—A memory that loses its contents when power supply is interrupted.

Westnet II—A proprietary local area network from Westinghouse Electric Co.

word—A group of one or more bytes.

XOR—The EXCLUSIVE-OR composite logic operation, which is the equivalent of one OR, two AND, and two NOT operations. Its truth table is identical to that of the OR operation except that it yields a false output when all inputs are true.

ZCL—Abbreviation for Zone Control Last state. It is a program instruction that is similar to MCR and holds the outputs within the zone in their last states.

zone—A portion of a ladder diagram isolated by the MCR and END MCR instructions, or by the ZCL and END ZCL instructions.

Bibliography

BECAUSE PROGRAMMABLE CONTROLLERS ARE BASED ON DIGITAL ELECTRONICS, A GOOD understanding of the latter will aid in learning about the former. A very good first primer on the subject of digital electronics, which includes material covered in the present work (e.g., number systems, codes, and Boolean functions) plus much more (e.g., flip-flops, shift registers, counters, displays, and hands-on projects) is

Hawkins, H.M. 1983. *Concepts of Digital Electronics*. Blue Ridge Summit: TAB BOOKS Inc.

In Chapter 10 of this work, it was noted that before a process can be controlled, the key variable (or variables) must be measured. The use of transducers to convert process variables into measurable voltage has received considerable study. The interested reader is referred to Chapter 5 of

Hallmark, C. 1973. *Electronics Measurements Simplified*. Blue Ridge Summit: TAB BOOKS Inc.

for a discussion of transducer measurements and instrumentation systems. The chapter discusses transducers for the measurement of displacement, force, acceleration, pressure, temperature, light, and radiation. A more complete treatment is given in

Kuecken, J. A. 1981. *How to Measure Anything with Electronic Instruments*. Blue Ridge Summit: TAB BOOKS Inc.

which includes frequency measurement, counting, ADC, and DAC in addition to the topics listed previously. And for the last word on this subject, see

Carr, J. J. 1987. *Digital Interfacing with an*
    Analog World 2nd Ed. Blue Ridge Summit:
TAB BOOKS Inc.

In order to capitalize fully on the power and flexibility of programmable controllers, a knowledge of process control theory is required. One of the best introductions to the subject is

Tedeschi, F. P. 1981. *How to Design, Build and Use*
    Electronic Control Systems. Blue Ridge Summit:
TAB BOOKS Inc.

This book provides the basics of both theory and applications. It includes topics not covered in the present work, such as Laplace transforms, Bode plots, and a general discussion on the stability of control systems. It makes no mention of programmable controllers: the control circuits presented are analog circuits. A more advanced book is

Coughanowr, D. R. and Koppel, L. B. 1965.
    Process Systems Analysis and Control.
New York: McGraw-Hill.

It provides an in-depth treatment of process control theory. Of course, it makes no mention of programmable controllers, since PCs were not invented when the book was written.

Throughout the present work, PCs have been compared with computers. Readers who are familiar with computers may wish to investigate

Weiss, D. 1987. *Microprocessors in Industrial*
    Measurement and Control. Blue Ridge Summit:
TAB BOOKS Inc.

and

Dalglish, R. L. 1987. *An Introduction to Control*
    and Measurement with Microcomputers. New York:
Cambridge University Press.

These books are very computer-oriented. For example, the latter spends six chapters on the internal operations of computers and on their structure, but only three chapters on interfacing with the outside world. It also assumes some knowledge of BASIC, Pascal, FORTRAN, or other high-level computer languages. Low-level languages such as ladder and Boolean are not mentioned. It was written primarily for computer-literate scientists and engineers.

Of course, there are books on programmable controllers. The standard, near-classic text

Jones, C. T. and Bryan, L. A. 1983. *Programmable Controllers: Concepts and Applications.* Atlanta: International Programmable Controls, Inc.

is an excellent book. Unfortunately, it is very difficult for the beginner, because it lacks much basic material. (The present work was written to remedy that deficiency.) The same holds true for

Wilhelm, Jr., R. E. 1985. *Programmable Controller Handbook.* Hasbrouck Heights: Hayden Book Co.

which is a massive work covering minute details of PC operations and applications. It is too difficult and overwhelming for the beginner, but it is an excellent reference source for the practicing technician or engineer. Other PC books include

Kissell, T. E. 1986. *Understanding and Using Programmable Controllers..* Englewood Cliffs: Prentice Hall.

Johannesson, G. 1985. *Programmable Control Systems.* Brookfield, VT: Brookfield Pub. Co.

Cox, R. A. 1984. *Technician's Guide to Programmable Controllers.* Albany: Delmar.

Bryan, E. A. and Bryan, L. A. 1987. *Programmable Controller Workbook and Study Guide.* Newchurch, K. Ed. Atlanta: International Programmable Controls, Inc.

Gilbert, R. 1986. *Programmable Controllers: Practices and Concepts.* Willow Grove, PA: Intertec Pub.

_____. 1987. *Programmable Controllers: Selected Applications Vol I..* Chicago and Atlanta: Industrial Text Co.

Flora, P. C., Ed. 1986. *International Programmable
Controllers Directory.* Blue Ridge Summit:
TAB BOOKS Inc.

Many of the books listed above, while too difficult for the beginner on his own, work
well in classroom situations. Wilhelm's book, for example, was written as a result of
his teaching experience in a community college. There are several courses offered on
PCs, ranging in length from two to five days, employing some of the texts mentioned
above. International Programmable Controls, Inc. offers one based on the Jones and Bryan
book (Industrial Programmable Controls, Inc., Attn: PC Training, 35 Glenlake Parkway-
Suite 445, Atlanta, GA 30328). *Engineer's Digest*, in cooperation with some domestic
PC suppliers, has in the past offered a course based on Gilbert's book (*Engineer's Digest*, Walker-Davis Publications, Inc., 2500 Office Center, Willow Grove, PA 19090).
The Instrument Society of America offers the T2000 series of courses on instrumentation
and control fundamentals, including PCs (Instrument Society of America, P.O. Box 12277,
Research Triangle Park, NC 27709). But be prepared to pay dearly for PC courses:
they range in price from $375 to more than $1000. They do offer the advantage of
classroom instruction and hands-on experience.

The PC world changes very rapidly. In order to stay on top of it, the three following
periodicals are recommended:

Control Engineering
Cahners Publishing Co.
Division of Reed Publishing USA
275 Washington St.
Newton, MA 02158

Chilton's I&CS
Chilton Co.
Chilton Way
Radnor, PA 19089

Engineer's Digest
Walker-Davis Publications, Inc.
2500 Office Center
Willow Grove, PA 19090

Index

Edited by Marilyn Johnson

DATE DUE